SpringerBriefs in Optimization

Series Editors

Panos M. Pardalos
János D. Pintér
Stephen M. Robinson
Tamás Terlaky
My T. Thai

SpringerBriefs in Optimization showcases algorithmic and theoretical techniques, case studies, and applications within the broad-based field of optimization. Manuscripts related to the ever-growing applications of optimization in applied mathematics, engineering, medicine, economics, and other applied sciences are encouraged.

For further volumes:
http://www.springer.com/series/8918

Alexander J. Zaslavski

Structure of Approximate Solutions of Optimal Control Problems

 Springer

Alexander J. Zaslavski
Department of Mathematics
Technion – Israel Institute of Technology
Haifa, Israel

ISSN 2190-8354   ISSN 2191-575X (electronic)
ISBN 978-3-319-01239-1   ISBN 978-3-319-01240-7 (eBook)
DOI 10.1007/978-3-319-01240-7
Springer New York Heidelberg Dordrecht London

Library of Congress Control Number: 2013943267

Mathematics Subject Classification: 49J15, 90C31, 54E35, 54E50, 54E52

Printed on acid-free paper

Springer is part of Springer Science+Business Media (www.springer.com)

Preface

In this book we study the structure of approximate solutions of optimal control problems considered on subintervals of a real line. We are interested in properties of approximate solutions which are independent of the length of the interval, for all sufficiently large intervals. The results in this book deal with the so-called turnpike property of the optimal control problems. To have this property means, roughly speaking, that the approximate solutions of the problems are determined mainly by the integrand (objective function) and are essentially independent of the choice of interval and endpoint conditions, except in regions close to the endpoints.

Turnpike properties are well known in mathematical economics. The term was first coined by P. Samuelson in 1948 when he showed that an efficient expanding economy would spend most of the time in the vicinity of a balanced equilibrium path (also called von Neumann path). Now it is well known that the turnpike property is a general phenomenon which holds for large classes of variational problems. For these classes of problems, using the Baire category approach, it was shown that the turnpike property holds for a generic (typical) problem.

In this book we generalize this result for a general class of optimal control problems. More precisely, in Chap. 2 of this book we consider a class of optimal control problems (with the same system of differential equations, the same functional constraints, and the same boundary conditions) which is identified with the corresponding complete metric space of objective functions (integrands). The main results of Chap. 2 establish the turnpike property for any element of a set which is a countable intersection of open everywhere dense sets in the space of integrands. This means that the turnpike property holds for most optimal control problems (integrands). In Chap. 3 we study infinite horizon optimal control problems corresponding to the space of integrands introduced in Chap. 2. A class of linear control problems is considered in Chap. 4.

Haifa, Israel   Alexander J. Zaslavski

Contents

1 Introduction 1
   1.1 Infinite Horizon Variational Problems 1
   1.2 The Turnpike Phenomenon 5
   1.3 Structure of Solutions of Variational Problems 7

2 Turnpike Properties of Optimal Control Problems 11
   2.1 Preliminaries 11
   2.2 The Main Results 13
   2.3 Uniform Boundedness of Trajectory-Control Pairs 17
   2.4 Auxiliary Results 22
   2.5 Proofs of Theorems 2.1–2.3 and 2.5 68

3 Infinite Horizon Problems 77
   3.1 Existence of Optimal Solutions 77
   3.2 Auxiliary Results 82
   3.3 Proof of Proposition 3.3 85
   3.4 Proof of Proposition 3.4 89
   3.5 Proof of Theorem 3.6 91
   3.6 Proof of Theorem 3.7 97
   3.7 Proof of Theorem 3.8 100

4 Linear Control Systems 105
   4.1 The Class of Problems 105
   4.2 Proof of Proposition 4.2 107
   4.3 A Continuity Property 111
   4.4 A Boundedness Property 116
   4.5 The Existence and Structure of Solutions 118

References 125

Index 127

Contents

References

Chapter 1
Introduction

1.1 Infinite Horizon Variational Problems

The study of optimal control problems and variational problems defined on infinite intervals and on sufficiently large intervals has been a rapidly growing area of research [4,5,8–13,16,18,22–24,27,34–38,40,45–47,50]. These problems arise in engineering [1,25,53], in models of economic growth [2,3,12,15,17,24,28,33,39, 41–43,50], in infinite discrete models of solid-state physics related to dislocations in one-dimensional crystals [7,44], in the calculus of variations on time scales [29,32] and in the theory of thermodynamical equilibrium for materials [14,26,30,31].

Consider the infinite horizon problem of minimizing the expression

$$\int_0^T f(t, x(t), x'(t))dt$$

as T grows to infinity where a function $x : [0, \infty) \to R^n$ is locally absolutely continuous (a. c.) and satisfies the initial condition $x(0) = x_0$, and f belongs to a complete metric space of functions to be described below.

We say that an a. c. function $x : [0, \infty) \to R^n$ is (f)-*overtaking optimal* if

$$\limsup_{T \to \infty} \int_0^T [f(t, x(t), x'(t)) - f(t, y(t), y'(t))]dt \leq 0$$

for any a. c. function $y : [0, \infty) \to R^n$ satisfying $y(0) = x(0)$.

This notion, known as the overtaking optimality criterion, was introduced in the economics literature [17, 43] and has been used in optimal control theory [12, 24, 49, 50].

Another type of optimality criterion for infinite horizon problems was introduced by Aubry and Le Daeron [7] in their study of the discrete Frenkel–Kontorova model related to dislocations in one-dimensional crystals. This optimality criterion was used in [14, 26, 30, 31, 44, 47, 50].

A.J. Zaslavski, *Structure of Approximate Solutions of Optimal Control Problems*,
SpringerBriefs in Optimization, DOI 10.1007/978-3-319-01240-7_1,
© Alexander J. Zaslavski 2013

Let I be either $[0, \infty)$ or $(-\infty, \infty)$. We say that an a. c. function $x : I \to R^n$ is an (f)-*minimal solution* if

$$\int_{T_1}^{T_2} f(t, x(t), x'(t))dt \leq \int_{T_1}^{T_2} f(t, y(t), y'(t))dt \leq 0$$

for each $T_1 \in I$, each $T_2 > T_1$, and each a. c. function $y : [T_1, T_2] \to R^n$ which satisfies $y(T_i) = x(T_i)$, $i = 1, 2$.

It is easy to see that every (f)-overtaking optimal function is an (f)-minimal solution.

In Chap. 1 of [50] and in [48] we considered a functional space of integrands $\mathcal{M}$ described below. We showed that for each $f \in \mathcal{M}$ and each $z \in R^n$ there exists a bounded (f)-minimal solution $Z : [0, \infty) \to R^n$ satisfying $Z(0) = z$ such that any other a. c. function $Y : [0, \infty) \to R^n$ is not "better" than Z. We also established that, given $f \in \mathcal{M}$ and a bounded set $E \subset R^n$, the $C([0, T])$ norms of approximate solutions $x : [0, T] \to R^n$ for the minimization problem on an interval $[0, T]$ with $x(0), x(T) \in E$ are bounded by some constant which depends only on f and E.

Let $a > 0$ be a constant and $\psi : [0, \infty) \to [0, \infty)$ be an increasing function such that $\psi(t) \to \infty$ as $t \to \infty$.

Denote by $|\cdot|$ the Euclidean norm in the n-dimensional Euclidean space R^n and denote by $\mathcal{M}$ the set of all continuous functions $f : [0, \infty) \times R^n \times R^n \to R^1$ which satisfy the following assumptions:

A(i)   For each $(t, x) \in [0, \infty) \times R^n$ the function $f(t, x, \cdot) : R^n \to R^1$ is convex;

A(ii)   The function f is bounded on $[0, \infty) \times E$ for any bounded set $E \subset R^n \times R^n$;

A(iii)   For each $(t, x, u) \in [0, \infty) \times R^n \times R^n$,

$$f(t, x, u) \geq \max\{\psi(|x|), \psi(|u|)|u|\} - a;$$

A(iv)   For each $M, \epsilon > 0$ there exist $\Gamma, \delta > 0$ such that

$$|f(t, x_1, u_1) - f(t, x_2, u_2)| \leq \epsilon \max\{f(t, x_1, u_1), f(t, x_2, u_2)\}$$

for each $t \in [0, \infty)$, each $u_1, u_2 \in R^n$ and each $x_1, x_2 \in R^n$ which satisfy

$$|x_i| \leq M, \ |u_i| \geq \Gamma, \ i = 1, 2, \quad \max\{|x_1 - x_2|, |u_1 - u_2|\} \leq \delta;$$

A(v)   For each $M, \epsilon > 0$ there exists $\delta > 0$ such that

$$|f(t, x_1, u_1) - f(t, x_2, u_2)| \leq \epsilon$$

for each $t \in [0, \infty)$, each $u_1, u_2 \in R^n$ and each $x_1, x_2 \in R^n$ which satisfy

$$|x_i|, |u_i| \leq M, \ i = 1, 2, \quad \max\{|x_1 - x_2|, |u_1 - u_2|\} \leq \delta.$$

Assumptions A(i)–A(v) were discussed in [50] with examples of integrands for which these assumptions hold. Note that assumption A(iii) implies that the function f grows to infinity as $|x| \to \infty$ and grows superlinearly with respect to u while assumption A(iv) means the uniform continuity of the function with respect to x and u on bounded sets.

It is an elementary exercise to show that an integrand $f = f(t, x, u) \in C^1([0, \infty) \times R^n \times R^n)$ belongs to $\mathcal{M}$ if f satisfies assumptions A(i), A(iii),

$$\sup\{|f(t, 0, 0)| : t \in [0, \infty)\} < \infty$$

and there exists an increasing function $\psi_0 : [0, \infty) \to [0, \infty)$ such that

$$\sup\{|\partial f / \partial x(t, x, u)|, \ |\partial f / \partial u(t, x, u)|\} \leq \psi_0(|x|)(1 + \psi(|u|)|u|)$$

for each $t \in [0, \infty)$, $x, u \in R^n$.

Therefore the space $\mathcal{M}$ contains many functions.

We equip the set $\mathcal{M}$ with the uniformity which is determined by the following base:

$$E(N, \epsilon, \lambda) = \{(f, g) \in \mathcal{M} \times \mathcal{M} : |f(t, x, u) - g(t, x, u)| \leq \epsilon$$

for each $t \in [0, \infty)$, each $u \in R^n$, and each $x \in R^n$ satisfying $|x|, |u| \leq N\}$

$$\cap \{(f, g) \in \mathcal{M} \times \mathcal{M} : (|f(t, x, u)| + 1)(|g(t, x, u)| + 1)^{-1} \in [\lambda^{-1}, \lambda]$$

for each $t \in [0, \infty)$, each $u \in R^n$, and each $x \in R^n$ satisfying $|x| \leq N\}$

where $N > 0$, $\epsilon > 0$, $\lambda > 1$.

Clearly, the uniform space $\mathcal{M}$ is Hausdorff and has a countable base. Therefore $\mathcal{M}$ is metrizable. We showed in Sect. 1.3 of [50] that the uniform space $\mathcal{M}$ is complete.

Put

$$I^f(T_1, T_2, x) = \int_{T_1}^{T_2} f(t, x(t), x'(t)) dt$$

where $f \in \mathcal{M}$, $0 \leq T_1 < T_2 < \infty$ and $x : [T_1, T_2] \to R^n$ is an a. c. function.

For $f \in \mathcal{M}$, $a, b \in R^n$ and numbers T_1, T_2 satisfying $0 \leq T_1 < T_2$, put

$$U^f(T_1, T_2, a, b) = \inf\{I^f(T_1, T_2, x) : x : [T_1, T_2] \to R^n$$

is an a. c. function satisfying $x(T_1) = a$, $x(T_2) = b\}$,

$$\sigma^f(T_1, T_2, a) = \inf\{U^f(T_1, T_2, a, b) : b \in R^n\}.$$

It is easy to see that $-\infty < U^f(T_1, T_2, a, b) < \infty$ for each $f \in \mathcal{M}$, each $a, b \in R^n$ and each pair of numbers T_1, T_2 satisfying $0 \leq T_1 < T_2$.

Let $f \in \mathcal{M}$. We say that an a. c. function $x : [0, \infty) \to R^n$ is an (f)-*good function* if for any a. c. function $y : [0, \infty) \to R^n$,

$$\inf\{I^f(0, T, y) - I^f(0, T, x) : T \in (0, \infty)\} > -\infty.$$

In Chap. 1 of [50] and in [48] we studied the set of (f)-good functions and proved the following results.

Theorem 1.1. *For each $h \in \mathcal{M}$ and each $z \in R^n$ there exists an (h)-good function $Z^h : [0, \infty) \to R^n$ satisfying $Z^h(0) = z$ such that:*

1. *For each $f \in \mathcal{M}$, each $z \in R^n$ and each a. c. function $y : [0, \infty) \to R^n$ one of the following properties holds:*

    (i) $I^f(0, T, y) - I^f(0, T, Z^f) \to \infty$ *as $T \to \infty$;*
    (ii) $\sup\{|I^f(0, T, y) - I^f(0, T, Z^f)| : T \in (0, \infty)\} < \infty,$

$$\sup\{|y(t)| : t \in [0, \infty)\} < \infty.$$

2. *For each $f \in \mathcal{M}$ and each number $M > 0$ there exist a neighborhood U of f in $\mathcal{M}$ and a number $Q > 0$ such that*

$$\sup\{|Z^g(t)| : t \in [0, \infty)\} \leq Q$$

    for each $g \in U$ and each $z \in R^n$ satisfying $|z| \leq M$.
3. *For each $f \in \mathcal{M}$ and each number $M > 0$ there exist a neighborhood U of f in $\mathcal{M}$ and a number $Q > 0$ such that for each $g \in U$, each $z \in R^n$ satisfying $|z| \leq M$, each $T_1 \geq 0$, $T_2 > T_1$ and each a. c. function $y : [T_1, T_2] \to R^n$ satisfying $|y(T_1)| \leq M$ the following relation holds:*

$$I^g(T_1, T_2, Z^g) \leq I^g(T_1, T_2, y) + Q.$$

4. *For each $f \in \mathcal{M}$ and each $z \in R^n$ the function $Z^f : [0, \infty) \to R^n$ is an (f)-minimal solution.*

Corollary 1.2. *Let $f \in \mathcal{M}$, $z \in R^n$ and let $y : [0, \infty) \to R^n$ be an a. c. function. Then y is an (f)-good function if and only if condition (ii) of Assertion 1 of Theorem 1.1 holds.*

Theorem 1.3. *For each $f \in \mathcal{M}$ there exist a neighborhood U of f in $\mathcal{M}$ and a number $M > 0$ such that for each $g \in U$ and each (g)-good function $x : [0, \infty) \to R^n$,*

$$\limsup_{t \to \infty} |x(t)| < M.$$

Our next result, which was also proved in Chap. 1 of [50] and in [48], shows that for every bounded set $E \subset R^n$ the $C([0, T])$ norms of approximate solutions $x :$

$[0, T] \to R^n$ for the minimization problem on an interval $[0, T]$ with $x(0), x(T) \in E$, are bounded by some constant which does not depend on T.

Theorem 1.4. *Let $f \in \mathcal{M}$ and M_1, M_2, c be positive numbers. Then there exist a neighborhood U of f in $\mathcal{M}$ and a number $S > 0$ such that for each $g \in U$, each $T_1 \in [0, \infty)$ and each $T_2 \in [T_1 + c, \infty)$ the following properties hold:*

(i) If $x, y \in R^n$ satisfy $|x|, |y| \leq M_1$ and if an a. c. function $v : [T_1, T_2] \to R^n$ satisfies

$$v(T_1) = x, \ v(T_2) = y, \ I^g(T_1, T_2, v) \leq U^g(T_1, T_2, x, y) + M_2,$$

then

$$|v(t)| \leq S, \ t \in [T_1, T_2].$$

(ii) If $x \in R^n$ satisfies $|x| \leq M_1$ and if an a. c. function $v : [T_1, T_2] \to R^n$ satisfies

$$v(T_1) = x, \ I^g(T_1, T_2, v) \leq \sigma^g(T_1, T_2, x) + M_2,$$

then

$$|v(t)| \leq S, \ t \in [T_1, T_2].$$

The results presented in this section are important ingredients in the proofs of turnpike results for variational problems [47, 48, 50].

1.2 The Turnpike Phenomenon

In Chap. 2 of [50] and in [49] we studied the structure of approximate solutions of the variational problems

$$\int_{T_1}^{T_2} f(t, z(t), z'(t))dt \to \min, \ z(T_1) = x, \ z(T_2) = y, \tag{P}$$

$$z : [T_1, T_2] \to R^n \text{ is an absolutely continuous function,}$$

where $T_1 \geq 0, T_2 > T_1, x, y \in R^n$, and $f : [0, \infty) \times R^n \times R^n \to R^1$ belongs to the complete metric space of integrands $\mathcal{M}$ which was introduced in Sect. 1.1.

Let $T_1 \geq 0, T_2 > T_1, x, y \in R^n, f : [0, \infty) \times R^n \times R^n \to R^1$ be an integrand and let δ be a positive number. We say that an absolutely continuous (a. c.) function $u : [T_1, T_2] \to R^n$ satisfying $u(T_1) = x, u(T_2) = y$ is a δ-*approximate* solution of the problem (P) if

$$\int_{T_1}^{T_2} f(t, u(t), u'(t))dt \leq \int_{T_1}^{T_2} f(t, z(t), z'(t))dt + \delta$$

for each a. c. function $z : [T_1, T_2] \to R^n$ satisfying $z(T_1) = x,\ z(T_2) = y$.

In Chap. 2 of [50] and in [49] we deal with the so-called *turnpike property* of the variational problems (P) associated with an integrand f. To have this property means that there exists a bounded continuous function $X_f : [0, \infty) \to R^n$ depending only on f such that for each pair of positive numbers $K, \epsilon > 0$ there exist positive constants $L = L(K, \epsilon)$ and $\delta = \delta(K, \epsilon)$ depending on ϵ, K such that if $u : [T_1, T_2] \to R^n$ is a δ-approximate solution of the problem (P) with

$$T_2 - T_1 \geq L,\ |u(T_i)| \leq K,\ i = 1, 2,$$

then

$$|u(t) - X_f(t)| \leq \epsilon \text{ for all } t \in [T_1 + \tau_1, T_2 - \tau_2],$$

where $\tau_1, \tau_2 \in [0, L]$.

If the integrand f possesses the turnpike property, then the solutions of variational problems with f are essentially independent of the choice of time interval and values at the endpoints except in regions close to the endpoints of the time interval. If a point t does not belong to these regions, then the value of a solution at t is close to a trajectory X_f ("turnpike") which is defined on the infinite time interval and depends only on f. This phenomenon has the following interpretation. If one wishes to reach a point A from a point B by a car in an optimal way, then one should turn to a turnpike, spend most of the time on it and then leave the turnpike to reach the required point.

Turnpike properties are well known in mathematical economics. The term was first coined by Samuelson in 1948 (see [42]) who showed that an efficient expanding economy would spend most of the time in the vicinity of a balanced equilibrium path (also called von Neumann path). This property was further investigated for optimal trajectories of models of economic dynamics (see, for example, [2, 15, 17, 28, 33, 41, 43, 50]). Many turnpike results are collected in [50].

In the classical turnpike theory the function f does not depend on the variable t, is strictly convex on the space $R^n \times R^n$, and satisfies a growth condition common in the literature. In this case, the turnpike property can be established, the turnpike X_f is a constant function and its value is a unique solution of the minimization problem $f(x, 0) \to \min, x \in R^n$.

It was shown in our research, which was summarized in [50], that the turnpike property is a general phenomenon which holds for large classes of variational problems without convexity assumptions. For these classes of problems a turnpike is not necessarily a constant function (singleton) but may instead be an absolutely continuous function on the interval $[0, \infty)$ as it was described above [49, 50].

More precisely, in Chap. 2 of [50] we studied the turnpike properties for variational problems with integrands which belong to the space $\mathcal{M}$ and showed that the turnpike property holds for a generic integrand $f \in \mathcal{M}$. Namely, we established the existence of a set $\mathcal{F} \subset \mathcal{M}$ which is a countable intersection of open everywhere dense sets in $\mathcal{M}$ such that each $f \in \mathcal{F}$ has the turnpike property. This result is presented and discussed in the next section.

1.3 Structure of Solutions of Variational Problems

Let $a > 0$ be a constant and let $\psi : [0, \infty) \to [0, \infty)$ be an increasing function such that

$$\psi(t) \to \infty \text{ as } t \to \infty.$$

We use the notation and definitions introduced in the previous sections. We consider the space of integrands $\mathcal{M}$ introduced in Sect. 1.1.

We equip the set $\mathcal{M}$ with two topologies where one is weaker than the other. We refer to them as the weak and the strong topologies, respectively. For the set $\mathcal{M}$ we consider the uniformity determined by the following base:

$$E_s(\epsilon) = \{(f, g) \in \mathcal{M} \times \mathcal{M} : |f(t, x, u) - g(t, x, u)| \leq \epsilon$$

$$\text{for each } t \in [0, \infty) \text{ and each } x, u \in R^n\},$$

where $\epsilon > 0$. It is not difficult to see that the uniform space $\mathcal{M}$ with this uniformity is metrizable and complete. This uniformity generates in $\mathcal{M}$ the strong topology.

We also equip the set $\mathcal{M}$ with the uniformity which is determined by the following base:

$$E(N, \epsilon, \lambda) = \{(f, g) \in \mathcal{M} \times \mathcal{M} : |f(t, x, u) - g(t, x, u)| \leq \epsilon$$

$$\text{for each } t \in [0, \infty) \text{ and each } x, u \in R^n \text{ satisfying } |x|, |u| \leq N,$$

$$(|f(t, x, u)| + 1)(|g(t, x, u)| + 1)^{-1} \in [\lambda^{-1}, \lambda]$$

$$\text{for each } t \in [0, \infty) \text{ and each } x, u \in R^n \text{ satisfying } |x| \leq N\},$$

where $N > 0$, $\epsilon > 0$, $\lambda > 1$. This uniformity which was introduced in Sect. 1.1 generates in $\mathcal{M}$ the weak topology.

In Chap. 2 of [50] and in [49] we established the existence of a set $\mathcal{F} \subset \mathcal{M}$ which is a countable intersection of open (in the weak topology) everywhere dense (in the strong topology) subsets of $\mathcal{M}$ such that the following theorems are valid.

Theorem 1.5.

1. *For each $g \in \mathcal{F}$ and each pair of (g)-good functions $v_i : [0, \infty) \to R^n$, $i = 1, 2$,*

$$|v_2(t) - v_1(t)| \to 0 \text{ as } t \to \infty.$$

2. *For each $g \in \mathcal{F}$ and each $y \in R^n$ there exists a (g)-overtaking optimal function $Y : [0, \infty) \to R^n$ satisfying $Y(0) = y$.*
3. *Let $g \in \mathcal{F}$, $\epsilon > 0$ and $Y : [0, \infty) \to R^n$ be a (g)-overtaking optimal function. Then there exists a neighborhood $\mathcal{U}$ of g in $\mathcal{M}$ with the weak topology such that the following property holds:*
   If $h \in \mathcal{U}$ and if $v : [0, \infty) \to R^n$ is an (h)-good function, then

$$|v(t) - Y(t)| \le \epsilon \text{ for all large } t.$$

Theorem 1.6. *Let $g \in \mathcal{F}$, $M, \epsilon > 0$ and let $Y : [0, \infty) \to R^n$ be a (g)-overtaking optimal function. Then there exists a neighborhood $\mathcal{U}$ of g in $\mathcal{M}$ with the weak topology and a number $\tau > 0$ such that for each $h \in \mathcal{U}$ and each (h)-overtaking optimal function $v : [0, \infty) \to R^n$ satisfying $|v(0)| \le M$,*

$$|v(t) - Y(t)| \le \epsilon \text{ for all } t \in [\tau, \infty).$$

Theorems 1.5 and 1.6 establish the existence of (g)-overtaking optimal functions and describe the asymptotic behavior of (g)-good functions for $g \in \mathcal{F}$.

Theorem 1.7. *Let $g \in \mathcal{F}$, $S_1, S_2, \epsilon > 0$ and let $Y : [0, \infty) \to R^n$ be a (g)-overtaking optimal function. Then there exists a neighborhood $\mathcal{U}$ of g in $\mathcal{M}$ with the weak topology, a number $L > 0$, and an integer $Q \ge 1$ such that if $h \in \mathcal{U}$, $T_1 \in [0, \infty)$, $T_2 \in [T_1 + LQ, \infty)$ and if an a. c. function $v : [T_1, T_2] \to R^n$ satisfies one of the following relations:*

(a) $|v(T_i)| \le S_1$, $i = 1, 2$, $\quad I^h(T_1, T_2, v) \le U^h(T_1, T_2, v(T_1), v(T_2)) + S_2$;

(b) $|v(T_1)| \le S_1$, $\quad I^h(T_1, T_2, v) \le \sigma^h(T_1, T_2, v(T_1)) + S_2$,

then the following property holds:
   There exist sequences of numbers $\{d_i\}_{i=1}^q$, $\{b_i\}_{i=1}^q \subset [T_1, T_2]$ such that

$$q \le Q, \ b_i < d_i \le b_i + L, \ i = 1, \dots, q,$$
$$|v(t) - Y(t)| \le \epsilon \text{ for each } t \in [T_1, T_2] \setminus \cup_{i=1}^q [b_i, d_i].$$

Theorem 1.8. *Let $g \in \mathcal{F}$, $S, \epsilon > 0$ and let $Y : [0, \infty) \to R^n$ be a (g)-overtaking optimal function. Then there exist a neighborhood $\mathcal{U}$ of g in $\mathcal{M}$ with the weak topology and numbers $\delta, L > 0$ such that for each $h \in \mathcal{U}$, each pair of numbers $T_1 \in [0, \infty)$, $T_2 \in [T_1 + 2L, \infty)$ and each a. c. function $v : [T_1, T_2] \to R^n$ which satisfies one of the following relations:*

(a) $|v(T_i)| \leq S$, $i = 1, 2$, $I^h(T_1, T_2, v) \leq U^h(T_1, T_2, v(T_1), v(T_2)) + \delta$;

(b) $|v(T_1)| \leq S$, $I^h(T_1, T_2, v) \leq \sigma^h(T_1, T_2, v(T_1)) + \delta$

the inequality $|v(t) - Y(t)| \leq \epsilon$ is valid for all $t \in [T_1 + L, T_2 - L]$.

Theorem 1.8 establishes the turnpike property for any $g \in \mathcal{F}$.

According to the results presented in this section, the turnpike property is a general phenomenon which holds for a large class of variational problems. For this class of problems, using the Baire category approach, it was shown that the turnpike property holds for a generic (typical) problem. Many results of this kind for other classes of variational problems are collected in [50]. Note that the generic approach of [50] is not limited to the turnpike property, but is also applicable to other problems in the optimization theory and nonlinear analysis [6, 51].

In this book we use the Baire category approach and generalize Theorem 1.8 for a general class of optimal control problems.

Our results are important for engineering and economic modeling. Optimal control problems studied in the book can be considered as a mathematical description of the corresponding continuous time models of economic growth without convexity assumptions which are usually present in the economic literature. Therefore the results of this book essentially enlarge the class of models which posses the turnpike property. In Chap. 4 of the book we apply our results to linear control systems which are very important in engineering where these systems are usually considered with quadratic cost functions. Here we again enlarge the class of linear control systems with the turnpike property considering nonconvex and nonautonomous integrands.

Chapter 2
Turnpike Properties of Optimal Control Problems

2.1 Preliminaries

Denote by $|\cdot|$ the Euclidean norm in the k-dimensional Euclidean space R^k. Let m, n be natural numbers.

In this chapter we study a control system described by a differential equation

$$x'(t) = G(t, x(t), u(t)) \text{ a. e. } t \in \mathcal{I}, \tag{2.1}$$

where $\mathcal{I}$ is either R^1 or $[T_1, \infty)$ or $[T_1, T_2]$ $(-\infty < T_1 < T_2 < \infty)$, and $x : \mathcal{I} \to R^n$ is an absolutely continuous (a. c.) function which satisfies

$$(t, x(t)) \in A \text{ for all } t \in \mathcal{I}, \tag{2.2}$$

where A is a subset of R^{n+1}. The control function $u : \mathcal{I} \to R^m$ is Lebesgue measurable and satisfies the feedback control constraints

$$u(t) \in U(t, x(t)) \text{ a. e. } t \in \mathcal{I}, \tag{2.3}$$

where $U : A \to 2^{R^m}$ is a point to set mapping with a graph

$$M = \{(t, x, u) : (t, x) \in A, \ u \in U(t, x)\}. \tag{2.4}$$

We suppose that M is a Borel measurable subset of R^{n+m+1} and that the function $G : M \to R^n$ is borelian.

For any $t \in R^1$ set

$$A(t) = \{x \in R^n : (t, x) \in A\}. \tag{2.5}$$

We assume that the set $A(t) \neq \emptyset$ for any $t \in R^1$.

A.J. Zaslavski, *Structure of Approximate Solutions of Optimal Control Problems*, SpringerBriefs in Optimization, DOI 10.1007/978-3-319-01240-7_2, © Alexander J. Zaslavski 2013

The performance of the above control system is described by an integral functional

$$I^f(T_1, T_2, x, u) = \int_{T_1}^{T_2} f(t, x(t), u(t))dt, \qquad (2.6)$$

where a borelian function $f : M \to R^1$ belongs to a complete metric space of functions $\mathfrak{M}$ defined below.

An a. c. function $x : \mathcal{I} \to R^n$, where $\mathcal{I}$ is either R^1 or $[T_1, \infty)$ or $[T_1, T_2]$ $(-\infty < T_1 < T_2 < \infty)$, will be called a *trajectory* if there exists a Lebesgue measurable function (referred to as a *control*) $u : \mathcal{I} \to R^m$ such that the pair (x, u) satisfies (2.1)–(2.3) and the function $t \to f(t, x(t), u(t))$ is locally Lebesgue integrable on $\mathcal{I}$.

For any $s \in R^1$ set $s_+ = \max\{s, 0\}$.

Let a_0 be a positive constant and let $\psi : [0, \infty) \to [0, \infty)$ be an increasing function such that

$$\psi(t) \to \infty \text{ as } t \to \infty.$$

Denote by $\mathfrak{M}$ the set of all borelian functions $f : M \to R^1$ which satisfy the following growth assumption:

(A)

$$f(t, x, u) \geq \max\{\psi(|x|),\ \psi(|u|),$$

$$\psi([|G(t, x, u)| - a_0|x|]_+)[|G(t, x, u)| - a_0|x|]_+\} - a_0$$

for each $(t, x, u) \in M$.

We equip the set $\mathfrak{M}$ with the uniformity which is determined by the following base:

$$E(N, \epsilon, \lambda) = \{(f, g) \in \mathfrak{M} \times \mathfrak{M} : |f(t, x, u) - g(t, x, u)| \leq \epsilon$$

$$\text{for each } (t, x, u) \in M \text{ satisfying } |x|, |u| \leq N\}$$

$$\cap \{(f, g) \in \mathfrak{M} \times \mathfrak{M} : (|f(t, x, u)| + 1)(|g(t, x, u)| + 1)^{-1} \in [\lambda^{-1}, \lambda]$$

$$\text{for each } (t, x, u) \in M \text{ satisfying } |x| \leq N\}, \qquad (2.7)$$

where $N > 0$, $\epsilon > 0$ and $\lambda > 1$.

Clearly, the uniform space $\mathfrak{M}$ is Hausdorff and has a countable base. Therefore $\mathfrak{M}$ is metrizable. It is not difficult to show that the uniform space $\mathfrak{M}$ is complete.

We consider functionals of the form $I^f(T_1, T_2, x, u)$ [see (2.6)], where $f \in \mathfrak{M}$, $-\infty < T_1 < T_2 < \infty$ and $x : [T_1, T_2] \to R^n$, $u : [T_1, T_2] \to R^m$ is a trajectory-control pair.

For $f \in \mathfrak{M}$, a pair of numbers $T_1 \in R^1$, $T_2 > T_1$ and (T_1, y), $(T_2, z) \in A$ set

$$U^f(T_1, T_2, y, z) = \inf\{I^f(T_1, T_2, x, u) : \ x : [T_1, T_2] \to R^n, \ u : [T_1, T_2] \to R^m$$

$$\text{is a trajectory-control pair satisfying } x(T_1) = y, \ x(T_2) = z\}, \qquad (2.8)$$

$$\sigma^f(T_1, T_2, y) = \inf\{U^f(T_1, T_2, y, h) : \ (T_2, h) \in A\}. \qquad (2.9)$$

Here we assume that the infimum over empty set is ∞.

2.2 The Main Results

Denote by $\mathfrak{M}_{reg}$ the set of all functions $f \in \mathfrak{M}$ which satisfy the following assumption:

(B) There exist a trajectory-control pair

$$x_f : R^1 \to R^n, \ u_f : R^1 \to R^m$$

and a number $b_f > 0$ such that:

(i)

$$U^f(T_1, T_2, x_f(T_1), x_f(T_2)) = I^f(T_1, T_2, x_f, u_f)$$

for each $T_1 \in R^1$ and each $T_2 > T_1$;

(ii)

$$\sup\{I^f(j, j+1, x_f, u_f) : j = 0, \pm 1, \pm 2, \ldots\} < \infty;$$

(iii) For each $S_1 > 0$ there exist $S_2 > 0$ and an integer $c > 0$ such that

$$I^f(T_1, T_2, x_f, u_f) \leq I^f(T_1, T_2, x, u) + S_2$$

for each $T_1 \in R^1$, each $T_2 \geq T_1 + c$, and each trajectory-control pair $x : [T_1, T_2] \to R^n$, $u : [T_1, T_2] \to R^m$ which satisfies $|x(T_1)|, |x(T_2)| \leq S_1$;

(iv) For each $\epsilon > 0$ there exists $\delta > 0$ such that for each $(T, z) \in A$ which satisfies

$$|z - x_f(T)| \leq \delta$$

there are

$$\tau_1 \in (T, T + b_f] \text{ and } \tau_2 \in [T - b_f, T),$$

and trajectory-control pairs

$$x_1 : [T, \tau_1] \to R^n, \ u_1 : [T, \tau_1] \to R^m,$$
$$x_2 : [\tau_2, T] \to R^n, \ u_2 : [\tau_2, T] \to R^m$$

which satisfy

$$x_1(T) = x_2(T) = z,$$
$$x_i(\tau_i) = x_f(\tau_i), \ i = 1, 2,$$
$$|x_1(t) - x_f(t)| \le \epsilon \text{ for all } t \in [T, \tau_1],$$
$$|x_2(t) - x_f(t)| \le \epsilon \text{ for all } t \in [\tau_2, T],$$
$$I^f(T, \tau_1, x_1, u_1) \le I^f(T, \tau_1, x_f, u_f) + \epsilon,$$
$$I^f(\tau_2, T, x_2, u_2) \le I^f(\tau_2, T, x_f, u_f) + \epsilon.$$

Note that assumption (B) means that the trajectory-control pair

$$x_f : R^1 \to R^n, \ u_f : R^1 \to R^m$$

is a solution of the corresponding infinite horizon optimal control problem associated with the integrand f, and that certain controllability properties hold near this trajectory-control pair.

In this chapter we will establish the following result.

Theorem 2.1.

1. *Let $f \in \mathfrak{M}_{reg}$ and $S_0 > 0$. Then there exists $S > 0$ such that for each pair of real numbers $T_1 < T_2$ and each trajectory-control pair*

$$x : [T_1, T_2] \to R^n, \ u : [T_1, T_2] \to R^m$$

which satisfies $|x(T_1)| \le S_0$ the following inequality holds:

$$I^f(T_1, T_2, x_f, u_f) \le I^f(T_1, T_2, x, u) + S.$$

2. *Let $f \in \mathfrak{M}_{reg}$. Then for each $s \in R^1$ and each trajectory-control pair*

$$x : [s, \infty) \to R^n, \ u : [s, \infty) \to R^m$$

one of the following relations holds:

(a)

$$I^f(s, t, x, u) - I^f(s, t, x_f, u_f) \to \infty \text{ as } t \to \infty;$$

(b)

$$sup\{|I^f(s, t, x_f, u_f) - I^f(s, t, x, u)| : \ t \in (s, \infty)\} < \infty.$$

Moreover, if the relation (b) holds, then

$$\sup\{|x(t)| : t \in [s, \infty)\} < \infty.$$

For each $f \in \mathfrak{M}_{reg}$ and each $r > 0$ we define a function $f_r \in \mathfrak{M}$ by

$$f_r(t, x, u) = f(t, x, u) + r\min\{|x - x_f(t)|, 1\} \text{ for all } (t, x, u) \in M.$$

It is easy to see that $f_r \in \mathfrak{M}_{reg}$ for each $f \in \mathfrak{M}_{reg}$ and each $r > 0$.

Let $\mathfrak{A}$ be a subset of $\mathfrak{M}_{reg}$ such that $f_r \in \mathfrak{A}$ for each $f \in \mathfrak{A}$ and each $r \in (0, 1)$. Denote by $\bar{\mathfrak{A}}$ the closure of $\mathfrak{A}$ in the uniform space $\mathfrak{M}$ and consider the topological subspace $\bar{\mathfrak{A}} \subset \mathfrak{M}$ with the relative topology.

In this chapter we will establish the existence of a set $\mathcal{F} \subset \bar{\mathfrak{A}}$ which is a countable intersection of open everywhere dense sets in $\bar{\mathfrak{A}}$ and such the following theorems hold.

Theorem 2.2. *For each $f \in \mathcal{F}$ and each $S > 0$ there exist a neighborhood U of f in $\mathfrak{M}$ and positive numbers δ, Q such that the following assertions hold:*

$$\inf\{U^g(T_1, T_2, y_1, y_2) : (T_i, y_i) \in A, \ i = 1, 2\} < \infty$$

for each $g \in U$, each $T_1 \in R^1$ and each $T_2 > T_1$;

for each $g \in U$, each $T_1 \in R^1$, each $T_2 \geq T_1 + 1$ and each trajectory-control pair

$$x : [T_1, T_2] \to R^n, \ u : [T_1, T_2] \to R^m$$

which satisfies

$$I^g(T_1, T_2, x, u) \leq \inf\{U^g(T_1, T_2, y_1, y_2) : (T_i, y_i) \in A, \ i = 1, 2\} + S$$

and

$$I^g(T_1, T_2, x, u) \leq U^g(T_1, T_2, x(T_1), x(T_2)) + \delta$$

the following inequality holds:

$$|x(t)| \leq Q \text{ for all } t \in [T_1, T_2].$$

Theorem 2.2 establishes uniform boundedness of approximate solutions of optimal control problems.

The next theorem is our first turnpike result.

Theorem 2.3. *Let $f \in \mathcal{F}$. Then there exists a bounded continuous function $X_f : R^1 \to R^n$ such that the following property holds.*

For each $S, \epsilon > 0$ there exist a neighborhood $\mathcal{U}$ of f in $\mathfrak{M}$ and real numbers $\Delta > 0$, $\delta \in (0, \epsilon)$ such that for each $g \in \mathcal{U}$, each $T_1 \in R^1$, each $T_2 \geq T_1 + 2\Delta$ and each trajectory-control pair

$$x : [T_1, T_2] \to R^n, \ u : [T_1, T_2] \to R^m$$

which satisfies

$$I^g(T_1, T_2, x, u) \leq \inf\{U^g(T_1, T_2, y_1, y_2) : (T_i, y_i) \in A, \ i = 1, 2\} + S$$

and

$$I^g(T_1, T_2, x, u) \leq U^g(T_1, T_2, x(T_1), x(T_2)) + \delta$$

the following inequality holds:

$$|x(t) - X_f(t)| \leq \epsilon \text{ for all } t \in [T_1 + \Delta, T_2 - \Delta].$$

Moreover, if $|x(T_1) - X_f(T_1)| \leq \delta$, then

$$|x(t) - X_f(t)| \leq \epsilon \text{ for all } t \in [T_1, T_2 - \Delta]$$

and if $|x(T_2) - X_f(T_2)| \leq \delta$, then

$$|x(t) - X_f(t)| \leq \epsilon \text{ for all } t \in [T_1 + \Delta, T_2].$$

Corollary 2.4. *Assume that $f \in \mathcal{F}$, S, Δ are positive numbers and that*

$$x : R^1 \to R^n, \ u : R^1 \to R^m$$

is a trajectory-control pair such that

$$I^f(T_1, T_2, x, u) = U^f(T_1, T_2, x(T_1), x(T_2))$$

for each $T_1 \in R^1$ and each $T_1 > T_1$, and

$$I^f(T_1, T_2, x, u) \leq \inf\{U^f(T_1, T_2, y_1, y_2) : (T_i, y_i) \in A, \ i = 1, 2\} + S$$

for each $T_1 \in R^1$ and each $T_2 > T_1 + \Delta$. Then

$$x(t) = X_f(t) \text{ for all } t \in R^1.$$

The next theorem is our second turnpike result.

Theorem 2.5. *Let $f \in \mathcal{F}$, let a bounded continuous function $X_f : R^1 \to R^n$ be as guaranteed by Theorem 2.3. and let ϵ, M be a pair of positive numbers. Then there exist a neighborhood $\mathcal{U}$ of f in $\mathfrak{M}$, real numbers $l > 0$, $L > 0$ and a natural number p such that for each $g \in \mathcal{U}$, each $T_1 \in R^1$, each $T_2 \geq T_1 + L$, and each trajectory-control pair*

$$x : [T_1, T_2] \to R^n, \ u : [T_1, T_2] \to R^m$$

which satisfies

$$I^g(T_1, T_2, x, u) \leq \inf\{U^g(T_1, T_2, y_1, y_2) : (T_i, y_i) \in A, \ i = 1, 2\} + M$$

there exist finite sequences

$$\{a_i\}_{i=1}^q, \ \{b_i\}_{i=1}^q \subset [T_1, T_2],$$

where $q \leq p$ is a natural number, such that

$$a_i \leq b_i \leq a_i + l \text{ for all integers } i = 1, \ldots, q$$

and

$$|x(t) - X_f(t)| \leq \epsilon \text{ for all } t \in [T_1, T_2] \setminus \cup_{i=1}^q [a_i, b_i].$$

This chapter is organized as follows. In Sect. 2.3 we study uniform boundedness of trajectory-control pairs. Section 2.4 contains auxiliary results. Theorems 2.1–2.3 and 2.5 are proved in Sect. 2.5.

2.3 Uniform Boundedness of Trajectory-Control Pairs

Proposition 2.6. *Let M_0, M_1, τ_0 be positive numbers. Then there exists $M_2 > M_1$ such that for each $f \in \mathfrak{M}$, each $T_1 \in R^1$, each $T_2 \in (T_1, T_1 + \tau_0]$, and each trajectory-control pair*

$$x : [T_1, T_2] \to R^n, \ u : [T_1, T_2] \to R^m$$

which satisfies

$$\inf\{|x(t)| : \ t \in [T_1, T_2]\} \leq M_1,$$

$$I^f(T_1, T_2, x, u) \leq M_0 \qquad (2.10)$$

the following inequality holds:

$$|x(t)| \leq M_2 \text{ for all } t \in [T_1, T_2].\tag{2.11}$$

Proof. Fix

$$\delta \in (0, \ \min\{8^{-1}(1 + a_0^{-1}), \ 16^{-1}\tau_0\})\tag{2.12}$$

[recall a_0 in assumption (A)]. By assumption (A) there exist $h_0 > M_1 + 1$ such that

$$f(t, x, u) \geq 4(M_0 + a_0\tau_0)\delta^{-1}$$
$$\text{for each } (t, x, u) \in M \text{ satisfying } |x| \geq h_0\tag{2.13}$$

and $\gamma_0 > 0$ such that

$$f(t, x, u) \geq 8[|G(t, x, u)| - a_0|x|]_+ \text{ for each } (t, x, u) \in M$$
$$\text{satisfying } |G(t, x, u)| - a_0|x| \geq \gamma_0.\tag{2.14}$$

Choose a number

$$M_2 > 8(M_0 + \tau_0 a_0 + \gamma_0\delta + h_0) + 8M_1.\tag{2.15}$$

Let $f \in \mathfrak{M}, T_1 \in R^1, T_2 \in (T_1, T_1 + \tau_0]$ and

$$x : [T_1, T_2] \to R^n, \ u : [T_1, T_2] \to R^m$$

be a trajectory-control pair satisfying (2.10). We show that (2.11) holds.
Assume the contrary. Then there exists $t_0 \in [T_1, T_2]$ such that

$$|x(t_0)| > M_2.\tag{2.16}$$

By the choice of h_0, (2.10), (2.12), (2.13), and assumption (A), there exists $t_1 \in [T_1, T_2]$ such that

$$|x(t_1)| \leq h_0 \text{ and } |t_1 - t_0| \leq \delta.\tag{2.17}$$

There exists

$$t_2 \in [\min\{t_0, t_1\}, \ \max\{t_0, t_1\}]$$

such that

$$|x(t_2)| \geq |x(t)| \text{ for all } t \in [\min\{t_0, t_1\}, \ \max\{t_0, t_1\}].\tag{2.18}$$

We define

$$E_1 = \{t \in [\min\{t_1, t_2\}, \ \max\{t_1, t_2\}] :$$
$$|G(t, x(t), u(t))| \geq a_0|x(t)| + \gamma_0\},$$
$$E_2 = [\min\{t_1, t_2\}, \ \max\{t_1, t_2\}] \setminus E_1. \tag{2.19}$$

It follows from (2.1), (2.19), (2.17), (2.18), (2.14), (2.10), and assumption (A) that

$$|x(t_2) - x(t_1)| = |\int_{t_1}^{t_2} G(t, x(t), u(t))dt|$$

$$\leq a_0|\int_{t_1}^{t_2} |x(t)|dt| + \int_{E_1} (|G(t, x(t), u(t))| - a_0|x(t)|)_+ dt$$

$$+ \int_{E_2} (|G(t, x(t), u(t))| - a_0|x(t)|)_+ dt$$

$$\leq a_0|x(t_2)|\delta + \gamma_0\delta + 8^{-1} \int_{E_1} f(t, x(t), u(t))dt$$

$$\leq a_0|x(t_2)|\delta + \gamma_0\delta + 8^{-1}(M_0 + \tau_0 a_0).$$

Together with (2.12), (2.18), (2.16), and (2.17) this implies that

$$(7/8)M_2 - h_0 \leq (7/8)|x(t_2)| - |x(t_1)| \leq \gamma_0\delta + 8^{-1}(M_0 + \tau_0 a_0).$$

This contradicts (2.15). The contradiction we have reached proves Proposition 2.6.

$\square$

Proposition 2.6 and assumption (A) imply the following result.

Proposition 2.7. *Let $M_1 > 0$ and $0 < \tau_0 < \tau_1$. Then there exists $M_2 > 0$ such that for each $f \in \mathfrak{M}$, each $T_1 \in R^1$, each $T_2 \in [T_1 + \tau_0, T_1 + \tau_1]$ and each trajectory-control pair*

$$x : [T_1, T_2] \to R^n, \ u : [T_1, T_2] \to R^m$$

which satisfies

$$I^f(T_1, T_2, x, u) \leq M_1 \tag{2.20}$$

relation (2.11) holds.

Proposition 2.8. *Let $M_1, \epsilon > 0$ and $0 < \tau_0 < \tau_1$. Then there exists $\delta > 0$ such that for each $f \in \mathfrak{M}$, each $T_1 \in R^1$, each $T_2 \in [T_1 + \tau_0, T_1 + \tau_1]$, and each trajectory-control pair*

$$x : [T_1, T_2] \to R^n, \ u : [T_1, T_2] \to R^m$$

which satisfies (2.20) and each $t_1, t_2 \in [T_1, T_2]$ *satisfying* $|t_1 - t_2| \leq \delta$ *the inequality*

$$|x(t_1) - x(t_2)| \leq \epsilon$$

holds.

Proof. Let a number $M_2 > 0$ be as guaranteed in Proposition 2.7. By assumption (A) there exists $h_0 > 0$ such that

$$f(t, x, u) \geq 4\epsilon^{-1}(M_1 + a_0\tau_1 + 8)(|G(t, x, u)| - a_0|x|)_+$$

for each $(t, x, u) \in M$ satisfying $|G(t, x, u)| - a_0|x| \geq h_0$.   (2.21)

Fix a number

$$\delta \in (0, \epsilon(4a_0 M_2 + 4h_0 + 4)^{-1}).$$ (2.22)

Let $f \in \mathfrak{M}, T_1 \in R^1, T_2 \in [T_1 + \tau_0, T_1 + \tau_1]$,

$$x : [T_1, T_2] \to R^n, \ u : [T_1, T_2] \to R^m$$

be a trajectory-control pair satisfying (2.20) and let

$$t_1, \ t_2 \in [T_1, T_2], \ 0 < t_2 - t_1 \leq \delta.$$ (2.23)

Set

$$E_1 = \{t \in [t_1, \ t_2] : \ |G(t, x(t), u(t))| - a_0|x(t)| \geq h_0\},$$
$$E_2 = [t_1, \ t_2] \setminus E_1.$$ (2.24)

It follows from the choice of M_2, (2.20), (2.11), (2.23), (2.24), (2.21), (2.22), and (2.1) that

$$|x(t_2) - x(t_1)| \leq a_0 \int_{t_1}^{t_2} |x(t)|dt + \int_{t_1}^{t_2} (|G(t, x(t), u(t))| - a_0|x(t)|)_+ dt$$

$$\leq a_0\delta M_2 + \delta h_0 + \int_{E_1} (|G(t, x(t), u(t))| - a_0|x(t)|)dt$$

$$\leq a_0\delta M_2 + \delta h_0 + \epsilon(4(M_1 + a_0\tau_1 + 8))^{-1} \int_{E_1} f(t, x(t), u(t))dt$$

$$\leq a_0\delta M_2 + \delta h_0 + 4^{-1}\epsilon \leq \epsilon.$$

Proposition 2.8 is proved.   □

Proposition 2.9. *Let $f \in \mathfrak{M}$, $0 < c_1 < c_2$, and $D, \epsilon > 0$. Then there exists a neighborhood V of f in $\mathfrak{M}$ such that for each $g \in V$, each $T_1 \in R^1$, each $T_2 \in [T_1 + c_1, T_1 + c_2]$, and each trajectory-control pair*

$$x : [T_1, T_2] \to R^n, \ u : [T_1, T_2] \to R^m$$

which satisfies

$$\min\{I^f(T_1, T_2, x, u), \ I^g(T_1, T_2, x, u)\} \le D$$

the inequality

$$|I^f(T_1, T_2, x, u) - I^g(T_1, T_2, x, u)| \le \epsilon$$

holds.

Proof. By Proposition 2.7 there exists $S > 0$ such that for each $g \in \mathfrak{M}$, each $T_1 \in R^1$, each $T_2 \in [T_1 + c_1, T_1 + c_2]$, and each trajectory-control pair

$$x : [T_1, T_2] \to R^n, \ u : [T_1, T_2] \to R^m$$

which satisfies

$$I^g(T_1, T_2, x, u) \le D + 1$$

the following inequality holds:

$$|x(t)| \le S \text{ for all } t \in [T_1, T_2]. \tag{2.25}$$

There exist $\delta \in (0, 1)$, $N > S$ and $\Gamma > 1$ such that

$$\psi(N) \ge 4a_0 + 4, \ \delta c_2 \le 8^{-1}\epsilon,$$
$$(\Gamma - 1)(D + a_0 c_2 + c_2) \le 8^{-1}\epsilon. \tag{2.26}$$

Define

$$V = \{g \in \mathfrak{M} : \ (f, g) \in E(N, \delta, \Gamma)\}. \tag{2.27}$$

Assume that $g \in V$, each $T_1 \in R^1$, each $T_2 \in [T_1 + c_1, T_1 + c_2]$ and

$$x : [T_1, T_2] \to R^n, \ u : [T_1, T_2] \to R^m$$

is a trajectory-control pair which satisfies

$$\min\{I^f(T_1, T_2, x, u), \ I^g(T_1, T_2, x, u)\} \le D. \tag{2.28}$$

Set

$$E_1 = \{t \in [T_1, T_2] : |u(t)| \leq N\}, \ E_2 = [T_1, T_2] \setminus E_1. \qquad (2.29)$$

It follows from (2.28), (2.29), (2.27), and the definition of S that (2.25) holds and that

$$|f(t, x(t), u(t)) - g(t, x(t), u(t))| \leq \delta \text{ for all } t \in E_1. \qquad (2.30)$$

Define

$$h(t) = \min\{f(t, x(t), u(t)), \ g(t, x(t), u(t))\}$$

for all $t \in [T_1, T_2]$. By (2.25), (2.27), (2.29), (2.26), and assumption (A), for each $t \in E_2$,

$$(f(t, x(t), u(t)) + 1)(g(t, x(t), u(t)) + 1)^{-1} \in [\Gamma^{-1}, \Gamma]$$

and

$$|f(t, x(t), u(t)) - g(t, x(t), u(t))| \leq (\Gamma - 1)(h(t) + 1).$$

It follows from (2.29), (2.30), assumption (A), (2.28), (2.26) and the inequality above that

$$|I^f(T_1, T_2, x, u) - I^g(T_1, T_2, x, u)| \leq \delta c_2 + (\Gamma - 1) \int_{E_2} h(t)dt + (\Gamma - 1)c_2$$

$$\leq \delta c_2 + (\Gamma - 1)c_2 + (\Gamma - 1)(D + a_0 c_2) \leq \epsilon.$$

Proposition 2.9 is proved. $\square$

2.4   Auxiliary Results

Consider the control system described by (2.1)–(2.5). Suppose that $f \in \mathfrak{M}$ and that there exists a trajectory-control pair

$$x_* : R^1 \to R^n, \ u_* : R^1 \to R^m$$

such that:

(i)

$$U^f(T_1, T_2, x_*(T_1), x_*(T_2)) = I^f(T_1, T_2, x_*, u_*)$$

for each $T_1 \in R^1$ and each $T_2 > T_1$;

(ii)

$$\sup\{I^f(j, j+1, x_*, u_*) : j = 0, \pm 1, \pm 2, \dots\} < \infty;$$

(iii) For each $S_1 > 0$ there exist $S_2 > 0$ and an integer $c > 0$ such that

$$I^f(T_1, T_2, x_*, u_*) \leq I^f(T_1, T_2, x, u) + S_2$$

for each $T_1 \in R^1$, each $T_2 \geq T_1 + c$ and each trajectory-control pair

$$x : [T_1, T_2] \to R^n, \ u : [T_1, T_2] \to R^m$$

which satisfies $|x(T_1)|, \ |x(T_2)| \leq S_1$;

(iv) There exists $b_* > 0$ such that for each $\epsilon > 0$ there exists $\delta > 0$ such that for each $(T, z) \in A$ which satisfies

$$|z - x_*(T)| \leq \delta$$

there are

$$\tau_1 \in (T, T + b_*] \text{ and } \tau_2 \in [T - b_*, T),$$

and trajectory-control pairs

$$x_1 : [T, \tau_1] \to R^n, \ u_1 : [T, \tau_1] \to R^m,$$
$$x_2 : [\tau_2, T] \to R^n, \ u_2 : [\tau_2, T] \to R^m$$

which satisfy

$$x_1(T) = x_2(T) = z,$$
$$x_i(\tau_i) = x_*(\tau_i), \ i = 1, 2,$$
$$|x_1(t) - x_*(t)| \leq \epsilon \text{ for all } t \in [T, \tau_1],$$
$$|x_2(t) - x_*(t)| \leq \epsilon \text{ for all } t \in [\tau_2, T],$$
$$I^f(T, \tau_1, x_1, u_1) \leq I^f(T, \tau_1, x_*, u_*) + \epsilon,$$
$$I^f(\tau_2, T, x_2, u_2) \leq I^f(\tau_2, T, x_*, u_*) + \epsilon.$$

It follows from property (ii) and Proposition 2.7 that

$$\sup\{|x_*(t)| : \ t \in R^1\} < \infty. \tag{2.31}$$

Note that properties (i)–(v) mean that the trajectory-control pair

$$x_* : R^1 \to R^n, \ u_* : R^1 \to R^m$$

is a solution of the corresponding infinite horizon optimal control problem associated with the integrand f and that certain controllability properties hold near this trajectory-control pair.

Lemma 2.10. *Let $S_0 > 0$. Then there exists $S > 0$ and an integer $c \geq 1$ such that for each $T_1 \in R^1$, each $T_2 \geq T_1 + c$ and each trajectory-control pair*

$$x : [T_1, T_2] \to R^n, \ u : [T_1, T_2] \to R^m,$$

satisfying $|x(T_1)| \leq S_0$ the inequality

$$I^f(T_1, T_2, x_*, u_*) \leq I^f(T_1, T_2, x, u) + S \tag{2.32}$$

holds.

Proof. Fix a number S_2 which satisfies

$$S_2 > S_0,$$

$$\psi(S_2) > a_0 + \sup\{|I^f(j, j+1, x_*, u_*)| : \ j = 0, \pm 1, \pm 2, \dots\}. \tag{2.33}$$

By property (iii) there exists a number $S_1 > 0$ and an integer $c > 0$ such that for each $T_1 \in R^1$, each $T_2 \geq T_1 + c$, and each trajectory-control pair

$$x : [T_1, T_2] \to R^n, \ u : [T_1, T_2] \to R^m$$

which satisfies $|x(T_1)|, \ |x(T_2)| \leq S_2$ the inequality

$$I^f(T_1, T_2, x_*, u_*) \leq I^f(T_1, T_2, x, u) + S_1$$

holds.

Fix a number

$$S > S_1 + 2 + 2a_0(2 + c)$$

$$+(c + 4) \sup\{|I^f(j, j+1, x_*, u_*)| : \ j = 0, \pm 1, \pm 2, \dots\}. \tag{2.34}$$

Let $T_1 \in R^1, T_2 \geq T_1 + c$ and

$$x : [T_1, T_2] \to R^n, \ u : [T_1, T_2] \to R^m$$

be a trajectory-control pair satisfying $|x(T_1)| \leq S_0$. We show that (2.32) holds. By the choice of S, S_1, and c, we may assume that

$$|x(T_2)| > S_2. \tag{2.35}$$

Set

$$T_3 = \sup\{t \in [T_1, T_2] : |x(t)| \le S_2\}. \tag{2.36}$$

It follows from the choice of S_1, (2.36), (2.33), and assumption (A) that

$$I^f(T_1, T_3, x_*, u_*) - I^f(T_1, T_3, x, u)$$
$$\le S_1 + 2a_0(1 + c) + (c + 2)\sup\{|I^f(j, j + 1, x_*, u_*)| : j = 0, \pm 1, \pm 2, \ldots\}. \tag{2.37}$$

By assumption (A), (2.33) and (2.36),

$$I^f(T_3, T_2, x_*, u_*) - I^f(T_3, T_2, x, u)$$
$$\le (T_2 - T_3 + 2)\sup\{|I^f(j, j + 1, x_*, u_*)| : j = 0, \pm 1, \pm 2, \ldots\}$$
$$+ 2a_0 - (\psi(S_2) - a_0)(T_2 - T_3)$$
$$\le 2\sup\{|I^f(j, j + 1, x_*, u_*)| : j = 0, \pm 1, \pm 2, \ldots\}.$$

Together with (2.34) and (2.37) this implies (2.32). Lemma 2.10 is proved. □

Fix a number $d_0 > 0$ and define a continuous function $\phi : A \to R^1$ by

$$\phi(t, x) = \min\{|x - x_*(t)|, d_0\} \text{ for each } (t, x) \in A. \tag{2.38}$$

For any $r > 0$ we define $f_r \in \mathfrak{M}$ by

$$f_r(t, x, u) = f(t, x, u) + r\phi(t, x) \text{ for each } (t, x, u) \in M. \tag{2.39}$$

We have the following simple auxiliary result.

Lemma 2.11. *Let V be a neighborhood of f in $\mathfrak{M}$. Then there exists $r_0 > 0$ such that $f_r \in V$ for every $r \in (0, r_0)$.*

Fix $r \in (0, 1]$ and set

$$\Lambda_0 = \sup\{|I^f(j, j + 1, x_*, u_*)| : j = 0, \pm 1, \pm 2, \ldots\},$$
$$\Lambda_1 = \sup\{|x_*(t)| : t \in R^1\}. \tag{2.40}$$

Lemma 2.12. *Let $b_1 \ge 1$ be an integer and let $b_2, Q, S > 0$. Then there exists $S_1 > S$ such that*

$$|x(t)| \le S_1 \text{ for all } t \in [T_1, T_2] \tag{2.41}$$

for each $T_1 \in R^1$, each $T_2 \geq T_1 + 2b_1$, and each trajectory-control pair

$$x : [T_1, T_2] \to R^n, \ u : [T_1, T_2] \to R^m$$

which satisfies the following conditions:

(a)

$$|x(T_1)|, \ |x(T_2)| \leq S$$

and

$$I^{fr}(T_1, T_2, x, u) \leq U^{fr}(T_1, T_2, x(T_1), x(T_2)) + Q;$$

(b) There exist

$$\tau_1 \in (T_1, T_1 + b_1] \text{ and } \tau_2 \in [T_2 - b_1, T_2)$$

such that

$$U^{fr}(T_1, \tau_1, x(T_1), x_*(\tau_1)) \leq b_2,$$
$$U^{fr}(\tau_2, T_2, x_*(\tau_2), x(T_2)) \leq b_2. \tag{2.42}$$

Proof. Choose a number S_0 such that

$$S_0 > S + 1,$$

$$\psi(S_0) > a_0 + \sup\{|I^f(j, j+1, x_*, u_*)| : \ j = 0, \pm1, \pm2, \dots\} + 4. \tag{2.43}$$

By property (iii), there exist $Q_1 > 0$ and an integer $c_1 \geq 1$ such that for each $T_1 \in R^1$, each $T_2 \geq T_1 + c_1$, and each trajectory-control pair

$$x : [T_1, T_2] \to R^n, \ u : [T_1, T_2] \to R^m$$

which satisfies

$$|x(T_i)| \leq S_0 + 1, \ i = 1, 2$$

the following inequality holds:

$$I^f(T_1, T_2, x_*, u_*) \leq I^f(T_1, T_2, x, u) + Q_1. \tag{2.44}$$

Fix a number

$$\tilde{S} > 4 + Q + 2Q_1 + 2a_0(b_1 + c_1 + 3) + 2b_2 + 2$$

$$+ \Lambda_0[2(c_1 + 3) + 2\Lambda_0(c_1 + 3) + 2a_0(b_1 + c_1 + 3) + 2b_2 + 2 + Q + 2Q_1]. \tag{2.45}$$

By Proposition 2.6 there exists $S_1 > S_0+2$ such that for each $g \in \mathfrak{M}$, each $T_1 \in R^1$, each

$$T_2 \in (T_1, T_1 + Q + 2b_2 + 2 + 2a_0(b_1 + c_1 + 3) + 2Q_1 + 2(c_1 + 3)\Lambda_0]$$

and each trajectory-control pair

$$x : [T_1, T_2] \to R^n, \ u : [T_1, T_2] \to R^m$$

which satisfies

$$\inf\{|x(t)| : t \in [T_1, T_2]\} \le S_0,$$
$$I^g(T_1, T_2, x, u) \le \tilde{S} \tag{2.46}$$

the following inequality holds:

$$|x(t)| \le S_1 \text{ for all } t \in [T_1, T_2]. \tag{2.47}$$

Assume that $T_1 \in R^1$, $T_2 \ge T_1 + 2b_1$ and that a trajectory-control pair

$$x : [T_1, T_2] \to R^n, \ u : [T_1, T_2] \to R^m$$

satisfies conditions (a) and (b). Let τ_1 and τ_2 be as guaranteed in condition (b). We show that (2.41) holds.

Assume the contrary. Then there exists $T_0 \in [T_1, T_2]$ such that

$$|x(T_0)| > S_1. \tag{2.48}$$

By condition (b) there exists a trajectory-control pair

$$x_1 : [T_1, T_2] \to R^n, \ u_1 : [T_1, T_2] \to R^m$$

such that

$$x_1(T_j) = x(T_j), \ x_1(\tau_j) = x_*(\tau_j), \ j = 1, 2,$$
$$x_1(t) = x_*(t) \text{ and } u_1(t) = u_*(t) \text{ for all } t \in [\tau_1, \tau_2],$$
$$I^{f_r}(T_1, \tau_1, x_1, u_1) \le b_2 + 1,$$
$$I^{f_r}(\tau_2, T_2, x_1, u_1) \le b_2 + 1. \tag{2.49}$$

It follows from (2.49) conditions (a) and (b), and assumption (A) that

$$I^{f_r}(T_1, T_2, x, u) \le I^{f_r}(T_1, T_2, x_1, u_1) + Q$$
$$\le Q + I^{f_r}(T_1, T_2, x_*, u_*) + 2a_0b_1 + 2b_2 + 2. \tag{2.50}$$

Set

$$t_1 = \sup\{t \in [T_1, T_0] : |x(t)| \le S_0\},$$
$$t_2 = \inf\{t \in [T_0, T_2] : |x(t)| \le S_0\}. \tag{2.51}$$

It follows from the choice of Q_1 and c_1, (2.44), (2.51), (2.40), condition (a), and assumption (A) that

$$I^{f_r}(T_1, t_1, x_*, u_*) - I^{f_r}(T_1, t_1, x, u)$$
$$\le Q_1 + a_0(c_1 + 2) + (c_1 + 2)\Lambda_0,$$
$$I^{f_r}(t_2, T_2, x_*, u_*) - I^{f_r}(t_2, T_2, x, u)$$
$$\le Q_1 + a_0(c_1 + 2) + (c_1 + 2)\Lambda_0. \tag{2.52}$$

By (2.39), (2.40), (2.51), (2.43), and assumption (A),

$$I^{f_r}(t_1, t_2, x_*, u_*) - I^{f_r}(t_1, t_2, x, u)$$
$$\le (t_2 - t_1 + 2)\Lambda_0 + 2a_0 - (t_2 - t_1)(\psi(S_0) - a_0) \le 2\Lambda_0 + 2a_0 - 4(t_2 - t_1).$$

Together with (2.50) and (2.52) this implies that

$$t_2 - t_1$$
$$\le 2\Lambda_0 + 2a_0 + Q + 2a_0 b_1 + 2b_2 + 2 + 2Q_1 + 2a_0(c_1 + 2) + 2(c_1 + 2)\Lambda_0. \tag{2.53}$$

It follows from the choice of S_1, (2.53), (2.46), (2.47), (2.48), and (2.51) that

$$I^{f_r}(t_1, t_2, x, u) > \tilde{S}.$$

Together with (2.39), (2.50), (2.52), (2.40), and assumption (A) this implies that

$$-Q - 2a_0 b_1 - 2b_2 - 2$$
$$\le I^{f_r}(T_1, T_2, x_*, u_*) - I^{f_r}(T_1, T_2, x, u)$$
$$\le 2Q_1 + 2a_0(c_1 + 2) + 2(c_1 + 2)\Lambda_0$$
$$+ I^{f_r}(t_1, t_2, x_*, u_*) - I^{f_r}(t_1, t_2, x, u)$$
$$\le 2Q_1 + 2a_0(c_1 + 2) + 2(c_1 + 2)\Lambda_0 + \Lambda_0(t_2 - t_1 + 2) + 2a_0 - \tilde{S}.$$

By this relation and (2.53),

$$\tilde{S} \leq Q + 2Q_1 + 2a_0(b_1 + c_1 + 3) + 2(b_2 + 1)$$
$$+ \Lambda_0[2(c_1 + 3) + 2\Lambda_0(c_1 + 3) + 2a_0(b_1 + c_1 + 3) + 2b_2 + 2 + Q + 2Q_1].$$

This contradicts (2.45). The contradiction we have reached proves Lemma 2.12. □

Lemma 2.13. *Let $b_1 \geq 1$ be an integer and let $b_2, Q, S > 0$. Then there exists $S_0 > 0$ such that for each $T_1 \in R^1$, each $T_2 \geq T_1 + 2b_1$, and each trajectory-control pair*

$$x : [T_1, T_2] \to R^n, \ u : [T_1, T_2] \to R^m$$

which satisfies conditions (a) and (b) of Lemma 2.12 the following inequality holds:

$$I^{fr}(T, T+1, x, u) \leq S_0 \text{ for any } T \in [T_1, T_2 - 1]. \tag{2.54}$$

Proof. Let $S_1 > S$ be as guaranteed in Lemma 2.12. By property (iii), there exist $Q_1 > 0$ and an integer $c_1 \geq 1$ such that for each $T_1 \in R^1$, each $T_2 \geq T_1 + c_1$, and each trajectory-control pair

$$x : [T_1, T_2] \to R^n, \ u : [T_1, T_2] \to R^m$$

which satisfies

$$|x(T_i)| \leq S_1 + 1, \ i = 1, 2$$

the following inequality holds:

$$I^f(T_1, T_2, x_*, u_*) \leq I^f(T_1, T_2, x, u) + Q_1. \tag{2.55}$$

Fix a number

$$S_0 > 4 + Q + 2Q_1 + 2(b_2 + 1) + 2a_0(b_1 + c_1 + 4) + 2(c_1 + 4)\Lambda_0 \tag{2.56}$$

[recall a_0 in assumption (A)].

Assume that $T_1 \in R^1$, $T_2 \geq T_1 + 2b_1$ and that a trajectory-control pair

$$x : [T_1, T_2] \to R^n, \ u : [T_1, T_2] \to R^m$$

satisfies conditions (a) and (b) of Lemma 2.12.

Let τ_1 and τ_2 be as guaranteed in conditions (a) and (b) of Lemma 2.12. We show that (2.54) holds for any $T \in [T_1, T_2 - 1]$.

Assume the contrary. Then there exists $T \in [T_1, T_2 - 1]$ such that

$$I^{fr}(T, T+1, x, u) > S_0. \tag{2.57}$$

By condition (b) of Lemma 2.12, there exists a trajectory-control pair

$$x_1 : [T_1, T_2] \to R^n, \; u_1 : [T_1, T_2] \to R^m$$

which satisfies (2.49). By (2.49), conditions (a) and (b) of Lemma 2.12 and assumption (A), relation (2.50) is true. It follows from the choice of S_1 and Lemma 2.12 that

$$|x(t)| \leq S_1 \text{ for all } t \in [T_1, T_2] \tag{2.58}$$

In view of the choice of Q_1, c_1, (2.58), (2.40), and (2.39),

$$I^{f_r}(T_1, T, x_*, u_*) - I^{f_r}(T_1, T, x, u) \leq Q_1 + a_0(c_1 + 2) + (c_1 + 2)\Lambda_0,$$

$$I^{f_r}(T + 1, T_2, x_*, u_*) - I^{f_r}(T + 1, T_2, x, u) \leq Q_1 + a_0(c_1 + 2) + (c_1 + 2)\Lambda_0. \tag{2.59}$$

It follows from (2.50), (2.59), (2.57), (2.40), and assumption (A) that

$$-Q - 2a_0 b_1 - 2b_2 - 2$$

$$\leq I^{f_r}(T_1, T_2, x_*, u_*) - I^{f_r}(T_1, T_2, x, u)$$

$$\leq 2Q_1 + 2a_0(c_1 + 2) + 2(c_1 + 2)\Lambda_0$$

$$+ I^{f_r}(T, T + 1, x_*, u_*) - I^{f_r}(T, T + 1, x, u)$$

$$\leq 2Q_1 + 2a_0(c_1 + 4) + 2(c_1 + 4)\Lambda_0 - S_0.$$

This contradicts (2.56). The contradiction we have reached proves Lemma 2.13. □

Lemma 2.14. *Let $\Delta > 0$. Then there exists $\delta \in (0, \Delta)$ such that for each $T_1 \in R^1$, each $T_2 > T_1$, and each trajectory-control pair*

$$x : [T_1, T_2] \to R^n, \; u : [T_1, T_2] \to R^m$$

which satisfies

$$|x(T_i) - x_*(T_i)| \leq \delta, \; i = 1, 2 \tag{2.60}$$

the following inequality holds:

$$I^f(T_1, T_2, x, u) \geq I^f(T_1, T_2, x_*, u_*) - \Delta. \tag{2.61}$$

Proof. There exists $\delta \in (0, \Delta)$ such that property (iv) holds with $\epsilon = 3^{-1}\Delta$ (see the definition of f).

Assume that $T_1 \in R^1$, $T_2 > T_1$ and that a trajectory-control pair

$$x : [T_1, T_2] \to R^n, \ u : [T_1, T_2] \to R^m$$

satisfies (2.60). By (2.60), the choice of δ and property (iv), there exist

$$\tau_1 \in [T_1 - b_*, T_1) \text{ and } \tau_2 \in (T_2, T_2 + b_*]$$

and trajectory-control pairs

$$x_1 : [\tau_1, T_1] \to R^n, \ u_1 : [\tau_1, T_1] \to R^m$$

and

$$x_2 : [T_2, \tau_2] \to R^n, \ u_2 : [T_2, \tau_2] \to R^m$$

such that

$$x_i(T_i) = x(T_i), \ i = 1, 2,$$
$$x_i(\tau_i) = x_*(\tau_i), \ i = 1, 2,$$
$$I^f(\tau_1, T_1, x_1, u_1) \leq I^f(\tau_1, T_1, x_*, u_*) + 3^{-1}\Delta,$$
$$I^f(T_2, \tau_2, x_2, u_2) \leq I^f(T_2, \tau_2, x_*, u_*) + 3^{-1}\Delta. \tag{2.62}$$

It follows from the choice of (x_1, u_1), (x_2, u_2) and property (i) that

$$I^f(\tau_1, T_1, x_1, u_1) + I^f(T_1, T_2, x, u) + I^f(T_2, \tau_2, , x_2, u_2)$$
$$\geq I^f(\tau_1, \tau_2, x_*, u_*).$$

Together with (2.62) this implies (2.61). Lemma 2.14 is proved.   □

Lemma 2.15. *Let $\epsilon \in (0, d_0)$. Then there exist $\delta \in (0, \epsilon)$ and $\Delta > 0$ such that for each $T_1 \in R^1$, each $T_2 \geq T_1 + \Delta$, and each trajectory-control pair*

$$x : [T_1, T_2] \to R^n, \ u : [T_1, T_2] \to R^m$$

which satisfies

$$|x(T_i) - x_*(T_i)| \leq \delta, \ i = 1, 2,$$
$$I^{f_r}(T_1, T_2, x, u) \leq U^{f_r}(T_1, T_2, x(T_1), x(T_2)) + \delta \tag{2.63}$$

the following inequality holds:

$$|x(t) - x_*(t)| \leq \epsilon \text{ for all } t \in [T_1, T_2]. \tag{2.64}$$

Proof. There exists $\delta_0 \in (0, 8^{-1})$ such that property (iv) holds with $\epsilon = 1$ and $\delta = \delta_0$. There exists an integer b_1 such that

$$b_* + 1 < b_1 \le b_* + 2 \tag{2.65}$$

[recall b_* in the definition of f, property (iv)].

Set

$$Q = 1,$$
$$S = \sup\{|x_*(t)| : t \in R^1\} + 4,$$
$$b_2 = 2 + b_* + 2a_0 + (b_* + 3)\Lambda_0. \tag{2.66}$$

Let S_0 be as guaranteed in Lemma 2.13. By Proposition 2.8 there exists $\delta_1 \in (0, 8^{-1})$ such that for each $g \in \mathfrak{M}$, each $T \in R^1$, each trajectory-control pair

$$x : [T, T+1] \to R^n, \ u : [T, T+1] \to R^m$$

which satisfies

$$I^g(T, T+1, x, u) \le 1 + S_0 + 2\Lambda_0 + 2a_0 \tag{2.67}$$

and each $t_1, t_2 \in [T, T+1]$ satisfying $|t_1 - t_2| \le \delta_1$ the inequality

$$|x(t_1) - x(t_2)| \le 16^{-1}\epsilon$$

holds.

Choose a number

$$\epsilon_1 \in (0, \min\{8^{-1}, 8^{-1}\epsilon, 12^{-1}\epsilon r \delta_1 (4 + 4b_*)^{-1}\}). \tag{2.68}$$

There exists

$$\delta_2 \in (0, 4^{-1}\epsilon_1) \tag{2.69}$$

such that property (iv) (see the definition of f) holds with $\epsilon = \epsilon_1$, $\delta = \delta_2$. There exists

$$\delta \in (0, \delta_2) \tag{2.70}$$

such that Lemma 2.14 holds with $\Delta = 16^{-1}\epsilon\delta_1 r$. Fix a number

$$\Delta > 2b_* + 2b_1 + 4. \tag{2.71}$$

Assume that $T_1 \in R^1$, $T_2 \geq T_1 + \Delta$ and that a trajectory-control pair

$$x : [T_1, T_2] \to R^n, \ u : [T_1, T_2] \to R^m$$

satisfies (2.63). We show that (2.64) holds.

Assume the contrary. Then there exists a number t_1 such that

$$t_1 \in [T_1, T_2],$$
$$|x(t_1) - x_*(t_1)| > \epsilon. \tag{2.72}$$

By (2.63), property (iv), and the choice of δ_2 there exist

$$\tau_1 \in (T_1, T_1 + b_*] \text{ and } \tau_2 \in [T_2 - b_*, T_2) \tag{2.73}$$

and trajectory-control pairs

$$x_1 : [T_1, \tau_1] \to R^n, \ u_1 : [T_1, \tau_1] \to R^m,$$
$$x_2 : [\tau_2, T_2] \to R^n, \ u_2 : [\tau_2, T_2] \to R^m$$

which satisfy

$$x_i(T_i) = x(T_i), \ i = 1, 2,$$
$$x_i(\tau_i) = x_*(\tau_i), \ i = 1, 2,$$
$$|x_1(t) - x_*(t)| \leq \epsilon_1 \text{ for all } t \in [T_1, \tau_1],$$
$$|x_2(t) - x_*(t)| \leq \epsilon_1 \text{ for all } t \in [\tau_2, T_2],$$
$$I^f(T_1, \tau_1, x_1, u_1) \leq I^f(T_1, \tau_1, x_*, u_*) + \epsilon_1,$$
$$I^f(\tau_2, T_2, x_2, u_2) \leq I^f(\tau_2, T_2, x_*, u_*) + \epsilon_1. \tag{2.74}$$

Define a trajectory-control pair

$$x_3 : [T_1, T_2] \to R^n, \ u_3 : [T_1, T_2] \to R^m$$

by

$$x_3(t) = x_1(t), \ u_3(t) = u_1(t) \text{ for all } t \in [T_1, \tau_1],$$
$$x_3(t) = x_*(t), \ u_3(t) = u_*(t) \text{ for all } t \in (\tau_1, \tau_2],$$
$$x_3(t) = x_2(t), \ u_3(t) = u_2(t) \text{ for all } t \in (\tau_2, T_2].$$

The relations above, (2.63) and (2.74) imply that

$$I^{fr}(T_1, T_2, x, u) \le I^{fr}(T_1, T_2, x_3, u_3) + \delta. \tag{2.75}$$

It follows from (2.75), (2.39), (2.74), (2.73), (2.65), and the definition of x_3 that

$$I^{fr}(T_1, T_2, x_3, u_3) - I^{fr}(T_1, T_2, x_*, u_*)$$

$$\le \int_{T_1}^{\tau_1} \phi(t, x_1(t))dt + I^f(T_1, \tau_1, x_1, u_1)$$

$$-I^f(T_1, \tau_1, x_*, u_*) + \int_{\tau_2}^{T_2} \phi(t, x_2(t))dt$$

$$+I^f(\tau_2, T_2, x_2, u_2) - I^f(\tau_2, T_2, x_*, u_*)$$

$$\le 2\epsilon_1 + 2\epsilon_1 b_1.$$

Together with (2.75) and (2.39) this implies that

$$I^f(T_1, T_2, x_*, u_*) + 2\epsilon_1 + 2\epsilon_1 b_1 + \delta \ge I^{fr}(T_1, T_2, x, u)$$

$$\ge I^f(T_1, T_2, x, u) + r \int_{T_1}^{T_2} \phi(t, x(t))dt. \tag{2.76}$$

By (2.74), (2.73), (2.39), and (2.68),

$$U^{fr}(T_1, \tau_1, x(T_1), x_*(\tau_1)) \le I^f(T_1, \tau_1, x_*, u_*) + 1 + b_*,$$

$$U^{fr}(\tau_2, T_2, x_*(\tau_2), x(T_2)) \le I^f(\tau_2, T_2, x_*, u_*) + 1 + b_*. \tag{2.77}$$

There exists an interval

$$[d_1, d_2] \subset [T_1, T_2]$$

such that

$$d_2 - d_1 = 1,$$

$$t_1 \in [d_1, d_2]. \tag{2.78}$$

It follows from (2.63), (2.68)–(2.70), (2.73), and (2.77) that the trajectory-control pair

$$x : [T_1, T_2] \to R^n, \ u : [T_1, T_2] \to R^m$$

satisfies conditions (a) and (b) of Lemma 2.12 with b_1, b_2, Q, S defined by (2.65) and (2.66). Therefore by the choice of S_0 and Lemma 2.13,

$$I^{fr}(d_1, d_2, x, u) \le S_0. \tag{2.79}$$

It is easy to see that

$$I^{f_r}(d_1, d_2, x_*, u_*) \leq 2\Lambda_0 + 2a_0. \tag{2.80}$$

It follows from the choice of δ_1, (2.67), (2.78)–(2.80), and (2.72) that for every

$$t \in [d_1, d_2] \cap [t_1 - \delta_1, t_1 + \delta_1]$$

we have

$$|x(t) - x(t_1)| \leq 16^{-1}\epsilon,$$
$$|x_*(t) - x_*(t_1)| \leq 16^{-1}\epsilon$$

and

$$|x(t) - x_*(t)| \geq (3/4)\epsilon.$$

Therefore

$$r \int_{T_1}^{T_2} \phi(t, x(t))dt \geq (3/4)\epsilon r \delta_1$$

and in view of (2.76) and (2.68)–(2.70),

$$I^f(T_1, T_2, x, u) \leq I^f(T_1, T_2, x_*, u_*) - 4^{-1}\epsilon r \delta_1.$$

On the other hand in view of the choice of δ [see (2.70)], (2.63) and Lemma 2.14,

$$I^f(T_1, T_2, x, u) \geq I^f(T_1, T_2, x_*, u_*) - 16^{-1}\epsilon r \delta_1.$$

The contradiction we have reached proves Lemma 2.15.   □

Lemma 2.16. *Let $\epsilon \in (0, \min\{1, d_0\})$. Then there exist $\delta \in (0, \epsilon)$ and $\Delta_1 > 0$ such that for each $\Delta_2 > \Delta_1$ there exists a neighborhood $\mathcal{U}$ of f_r in $\mathfrak{M}$ for which the following property holds:*

For each $g \in \mathcal{U}$, each $T_1 \in R^1$, each $T_2 \in [T_1 + \Delta_1, T_1 + \Delta_2]$, and each trajectory-control pair

$$x : [T_1, T_2] \to R^n, \ u : [T_1, T_2] \to R^m$$

which satisfies

$$|x(T_i) - x_*(T_i)| \leq \delta, \ i = 1, 2,$$
$$I^g(T_1, T_2, x, u) \leq U^g(T_1, T_2, x(T_1), x(T_2)) + \delta \tag{2.81}$$

the following inequality holds:

$$|x(t) - x_*(t)| \leq \epsilon \text{ for all } t \in [T_1, T_2].$$ (2.82)

Proof. There exists $\delta_0 \in (0, 8^{-1})$ such that property (iv) (see the definition of f) holds with $\epsilon = 1$ and $\delta = \delta_0$. By Lemma 2.15 there exist

$$\delta_1 \in (0, \min\{4^{-1}\epsilon, 4^{-1}\delta_0\}),$$

$$\Delta_1 \geq 2b_* + 2$$ (2.83)

such that for each $T_1 \in R^1$, each $T_2 \geq T_1 + \Delta_1$, and each trajectory-control pair

$$x : [T_1, T_2] \to R^n, \ u : [T_1, T_2] \to R^m$$

which satisfies

$$|x(T_i) - x_*(T_i)| \leq \delta_1, \ i = 1, 2,$$
$$I^{f_r}(T_1, T_2, x, u) \leq U^{f_r}(T_1, T_2, x(T_1), x(T_2)) + \delta_1$$ (2.84)

relation (2.82) holds for all $t \in [T_1, T_2]$. Fix

$$\delta \in (0, 16^{-1}\delta_1).$$ (2.85)

Let $\Delta_2 > \Delta_1$. By Proposition 2.9 there exists a neighborhood $\mathcal{U}$ of f_r in $\mathfrak{M}$ such that for each $g \in \mathcal{U}$, each $T_1 \in R^1$, each $T_2 \in [T_1 + \Delta_1, T_1 + \Delta_2]$, and each trajectory-control pair

$$x : [T_1, T_2] \to R^n, \ u : [T_1, T_2] \to R^m$$

which satisfies

$$\min\{I^{f_r}(T_1, T_2, x, u), \ I^g(T_1, T_2, x, u)\} \leq 4 + 2a_0 + \Delta_2 + \Lambda_0(\Delta_2 + 2)$$ (2.86)

the inequality

$$|I^{f_r}(T_1, T_2, x, u) - I^g(T_1, T_2, x, u)| \leq \delta$$ (2.87)

holds.

Assume that

$$g \in \mathcal{U}, \ T_1 \in R^1, \ T_2 \in [T_1 + \Delta_1, T_1 + \Delta_2]$$

and a trajectory-control pair

$$x : [T_1, T_2] \to R^n, \ u : [T_1, T_2] \to R^m$$

satisfies (2.81). We show that (2.82) holds.

By (2.81), (2.83), property (iv), and the choice of δ_0 there exist

$$\tau_1 \in (T_1, T_1 + b_*] \text{ and } \tau_2 \in [T_2 - b_*, T_2) \qquad (2.88)$$

and trajectory-control pairs

$$x_1 : [T_1, \tau_1] \to R^n, \ u_1 : [T_1, \tau_1] \to R^m,$$
$$x_2 : [\tau_2, T_2] \to R^n, \ u_2 : [\tau_2, T_2] \to R^m$$

such that

$$x_i(T_i) = x(T_i), \ i = 1, 2,$$
$$x_i(\tau_i) = x_*(\tau_i), \ i = 1, 2,$$
$$|x_1(t) - x_*(t)| \le 1 \text{ for all } t \in [T_1, \tau_1],$$
$$|x_2(t) - x_*(t)| \le 1 \text{ for all } t \in [\tau_2, T_2],$$
$$I^f(T_1, \tau_1, x_1, u_1) \le I^f(T_1, \tau_1, x_*, u_*) + 1,$$
$$I^f(\tau_2, T_2, x_2, u_2) \le I^f(\tau_2, T_2, x_*, u_*) + 1. \qquad (2.89)$$

Define a trajectory-control pair

$$x_3 : [T_1, T_2] \to R^n, \ u_3 : [T_1, T_2] \to R^m$$

by

$$x_3(t) = x_1(t), \ u_3(t) = u_1(t) \text{ for all } t \in [T_1, \tau_1],$$
$$x_3(t) = x_*(t), \ u_3(t) = u_*(t) \text{ for all } t \in (\tau_1, \tau_2],$$
$$x_3(t) = x_2(t), \ u_3(t) = u_2(t) \text{ for all } t \in (\tau_2, T_2]. \qquad (2.90)$$

Relations (2.88), (2.89), (2.90), (2.39), and (2.40) imply that

$$U^{fr}(T_1, T_2, x(T_1), x(T_2)) \le I^{fr}(T_1, T_2, x_3, u_3)$$
$$\le r \int_{T_1}^{T_2} \phi(t, x_3(t)) dt + I^f(T_1, T_2, x_3, u_3)$$
$$\le 2b_* + 2 + I^f(T_1, T_2, x_*, u_*) \le 2b_* + 2 + \Lambda_0(\Delta_2 + 2) + 2a_0. \qquad (2.91)$$

It follows from (2.91), (2.83), (2.81), and the choice of $\mathcal{U}$ [see (2.86), (2.87)] that

$$|U^{f_r}(T_1, T_2, x(T_1), x(T_2)) - U^g(T_1, T_2, x(T_1), x(T_2))| \leq \delta$$

and

$$|I^{f_r}(T_1, T_2, x, u) - I^g(T_1, T_2, x, u)| \leq \delta.$$

Together with (2.81) and (2.85) this implies that

$$I^{f_r}(T_1, T_2, x, u) \leq U^{f_r}(T_1, T_2, x(T_1), x(T_2)) + 3\delta$$
$$\leq U^{f_r}(T_1, T_2, x(T_1), x(T_2)) + \delta_1.$$

Now (2.82) follows from the relation above, (2.81), (2.85) and the choice of δ_1, Δ_1 [see (2.83), (2.84)]. This completes the proof of Lemma 2.16. $\square$

Lemma 2.17. *Let $S > 0$ and $\delta \in (0, \min\{1, d_0\})$. Then there exist $\Delta \geq 1$ and a neighborhood $\mathcal{U}$ of f_r in $\mathfrak{M}$ such that for each $g \in \mathcal{U}$, each $T_1 \in R^1$, each $T_2 \geq T_1 + \Delta$ and each trajectory-control pair*

$$x : [T_1, T_2] \to R^n, \ u : [T_1, T_2] \to R^m$$

which satisfies

$$|x(t) - x_*(t)| \geq \delta \text{ for all } t \in [T_1, T_2] \tag{2.92}$$

the following inequality holds:

$$I^g(T_1, T_2, x, u) > I^g(T_1, T_2, x_*, u_*) + S. \tag{2.93}$$

Proof. Fix a number

$$D_1 > 16(\Lambda_0 + a_0 + b_* + S + 22). \tag{2.94}$$

By Proposition 2.7 there exists $D_2 > D_1$ such that for each $g \in \mathfrak{M}$, each $T_1 \in R^1$, each $T_2 \in [T_1 + 8^{-1}, T_1 + 8]$, and each trajectory-control pair

$$x : [T_1, T_2] \to R^n, \ u : [T_1, T_2] \to R^m$$

which satisfies

$$I^g(T_1, T_2, x, u) \leq 2D_1 + 4 \tag{2.95}$$

we have

$$|x(t)| \leq D_2 \text{ for all } t \in [T_1, T_2].$$

By Lemma 2.10 there exist $D_3 > D_2$ and an integer $c_1 \geq 1$ such that for each $T_1 \in R^1$, each $T_2 \geq T_1 + c_1$, and each trajectory-control pair

$$x : [T_1, T_2] \to R^n, \ u : [T_1, T_2] \to R^m$$

satisfying $|x(T_1)| \leq 2D_2 + 1$ the inequality

$$I^f(T_1, T_2, x_*, u_*) \leq I^f(T_1, T_2, x, u) + D_3 \tag{2.96}$$

holds.

Choose numbers

$$S_1 > 8(D_3 + 20 + c_1 + 4(b_* + 4 + c_1)(\Lambda_0 + a_0) + S),$$
$$D_4 > 8S_1(2 + 4a_0 + 4\Lambda_0). \tag{2.97}$$

By Proposition 2.9 there exists a neighborhood $\mathcal{U}$ of f_r in $\mathfrak{M}$ such that for each $g \in \mathcal{U}$, each $T_1 \in R^1$, each $T_2 \in [T_1 + 4^{-1}, T_1 + 8]$, and each trajectory-control pair

$$x : [T_1, T_2] \to R^n, \ u : [T_1, T_2] \to R^m$$

which satisfies

$$\min\{I^{f_r}(T_1, T_2, x, u), \ I^g(T_1, T_2, x, u)\} \leq 2D_4 + 4 \tag{2.98}$$

the inequality

$$|I^g(T_1, T_2, x, u) - I^{f_r}(T_1, T_2, x, u)| \leq 64^{-1} r\delta \tag{2.99}$$

holds.

Fix a number

$$\Delta > 64(r\delta)^{-1}[2S_1 + (c_1 + 2)(a_0 + \Lambda_0)]. \tag{2.100}$$

Assume that $g \in \mathcal{U}$, $T_1 \in R^1$, $T_2 \geq T_1 + \Delta$, and that a trajectory-control pair

$$x : [T_1, T_2] \to R^n, \ u : [T_1, T_2] \to R^m$$

satisfies (2.92). By assumption (A), (2.40), (2.97), and the choice of $\mathcal{U}$ [see (2.98), (2.99)] for each $t_1, t_2 \in [T_1, T_2]$ satisfying $t_2 - t_1 \in [2^{-1}, 2]$

$$I^g(t_1, t_2, x_*, u_*) \leq I^{f_r}(t_1, t_2, x_*, u_*) + 64^{-1}\delta \leq 3\Lambda_0 + 2a_0 + 1. \tag{2.101}$$

We show that (2.93) holds. There exists an integer p such that

$$p - 1 < T_2 - T_1 \leq p. \tag{2.102}$$

Set

$$b_0 = T_1, \ b_j = T_1 + j, \ j = 1, \ldots, p - 2, \ b_{p-1} = T_2, \ \tau_0 = T_1. \tag{2.103}$$

By induction we define a sequence $\{\tau_j\} \subset \{b_i\}_{i=0}^{p-1}$.

Assume that an integer $q \geq 0$, the sequence $\{\tau_i\}_{i=0}^q$ has been defined and that $\tau_q = b_{j(q)}$, where $0 \leq j(q) < p - 1$. If

$$I^g(b_{j(q)}, b_{j(q)+1}, x, u) \geq D_1, \tag{2.104}$$

then we set $\tau_{q+1} = b_{j(q)+1}$. If (2.104) does not hold and there exists an integer $k \in \{j(q) + 1, \ldots, p - 2\}$ such that

$$I^g(b_i, b_{i+1}, x, u) < D_4, \ i = j(q), \ldots, k - 1,$$
$$I^g(b_k, b_{k+1}, x, u) \geq D_4, \tag{2.105}$$

then $\tau_{q+1} = b_{k+1}$. Otherwise $\tau_{q+1} = b_{p-1}$.

Evidently, the construction of the sequence $\{\tau_i\}$ is completed in a finite number of steps. Let τ_Q be the last element of the sequence and let

$$q \in \{0, \ldots, Q - 1\}, \ j(q) \in \{0, \ldots, p - 2\}, \ \tau_q = b_{j(q)}. \tag{2.106}$$

We estimate

$$I^g(\tau_q, \tau_{q+1}, x, u) - I^g(\tau_q, \tau_{q+1}, x_*, u_*).$$

If (2.204) holds, then by (2.101) which holds with $t_1 = \tau_q$, $t_2 = \tau_{q+1}$ and (2.94),

$$I^g(\tau_q, \tau_{q+1}, x, u) - I^g(\tau_q, \tau_{q+1}, x_*, u_*) \geq (3/4)D_1. \tag{2.107}$$

Assume that (2.104) does not hold and there exists $k \in \{j(q)+1, \ldots, p-2\}$ which satisfies (2.105). We show that

$$I^g(\tau_q, \tau_{q+1}, x, u) - I^g(\tau_q, \tau_{q+1}, x_*, u_*) \geq (3/4)D_4. \tag{2.108}$$

It follows from (2.105), the choice of τ_{q+1}, (2.101), and (2.97) that

$$I^g(\tau_q, \tau_{q+1}, x, u) - I^g(\tau_q, \tau_{q+1}, x_*, u_*)$$
$$\geq I^g(b_{j(q)}, b_k, x, u) - I^g(b_{j(q)}, b_k, x_*, u_*) + (7/8)D_4. \tag{2.109}$$

By the choice of $\mathcal{U}$ [see (2.98), (2.99)], (2.105), (2.101), and (2.97) for $i = j(q), \ldots, k-1$,

$$|I^g(b_i, b_{i+1}, y, v) - I^{fr}(b_i, b_{i+1}, y, v)| \leq 64^{-1}\delta r,$$
$$(y, v) \in \{(x, u), (x_*, u_*)\}. \tag{2.110}$$

It follows from (2.38), (2.39), and (2.92) that

$$I^{fr}(b_{j(q)}, b_k, x, u) \geq I^f(b_{j(q)}, b_k, x, u) + r\delta(k - j(q)). \tag{2.111}$$

Since (2.104) does not hold it follows from the definition of D_2 [see (2.95)] that

$$|x(t)| \leq D_2 \text{ for all } t \in [b_{j(q)}, b_{j(q)+1}]. \tag{2.112}$$

There are two cases: (i) $k - j(q) \geq S_1$; (ii) $k - j(q) < S_1$.

Consider the case (i). By (2.112), the choice of D_3, c_1 [see (2.96)] and the inequality $k - j(q) \geq S_1$,

$$I^f(b_{j(q)}, b_k, x, u) \geq I^f(b_{j(q)}, b_k, x_*, u_*) - D_3. \tag{2.113}$$

Combining (2.109), (2.110), (2.111), (2.113), (2.97), and the inequality

$$k - j(q) \geq S_1$$

we obtain (2.108). If $k - j(q) < S_1$, then (2.108) follows from (2.109), (2.101), (2.97), and assumption (A).

Assume that (2.104) does not hold and there is no $k \in \{j(q) + 1, \ldots, p - 2\}$ satisfying (2.105). Then

$$\tau_{q+1} = b_{p-1}, \ q = Q - 1,$$
$$I^g(b_i, b_{i+1}, x, u) < D_4, \ i = j(q), \ldots, p - 2. \tag{2.114}$$

By the choice of $\mathcal{U}$ [see (2.98), (2.99)], (2.114), (2.101), and (2.97), relation (2.110) holds for $i = j(q), \ldots, p-2$. Relation (2.110), which holds for $i = j(q), \ldots, p-2$, (2.92) and (2.114) imply that

$$I^g(\tau_q, \tau_{q+1}, x, u) - I^g(\tau_q, \tau_{q+1}, x_*, u_*)$$
$$\geq I^g(\tau_q, \tau_{q+1}, x, u) - I^g(\tau_q, \tau_{q+1}, x_*, u_*) + r(\delta - 32^{-1}\delta)(T_2 - \tau_q). \tag{2.115}$$

Since (2.104) does not hold it follows from the choice of D_2 [see (2.95)] that

$$|x(t)| \leq D_2 \text{ for all } t \in [\tau_q, \tau_q + 1].$$

By this relation, the choice of D_3, c_1 [see (2.96)] and assumption (A),

$$I^f(\tau_q, T_2, x, u) - I^f(\tau_q, T_2, x_*, u_*)$$
$$\geq -D_3 - a_0 c_1 - \Lambda_0(c_1 + 2) - 2a_0.$$

Together with (2.115) this implies that

$$I^g(\tau_q, T_2, x, u) - I^g(\tau_q, T_2, x_*, u_*)$$
$$\geq 2^{-1} r\delta(T_2 - \tau_q) - D_3 - (a_0 + \Lambda_0)(c_1 + 2) \qquad (2.116)$$

(here $q = Q - 1$ [see (2.114)]).

We showed that if (2.104) holds, then (2.107) is valid, if (2.104) does not hold and there is $k \in \{j(q) + 1, \ldots, p - 2\}$ satisfying (2.105), then (2.108) holds; otherwise $q = Q - 1$ and (2.116) holds.

We show that (2.93) holds. For any integer $q \in \{0, \ldots, Q\}$ there is an integer $j(q) \in \{0, \ldots, p - 1\}$ such that $\tau_q = b_{j(q)}$. We may assume that for $q = Q - 1$ (2.104) does not hold and there is not an integer

$$k \in \{j(Q - 1) + 1, \ldots, p - 2\}$$

satisfying (2.105) with $q = Q - 1$. If $T_2 - \tau_{Q-1} \geq 8^{-1}\Delta$, then (2.93) follows from (2.100), (2.116), which holds with $q = Q - 1$, (2.107), (2.108), and (2.94). Therefore we may assume that

$$T_2 - \tau_{Q-1} < 8^{-1}\Delta.$$

This implies that

$$(7/8)\Delta \leq \tau_{Q-1} - T_1. \qquad (2.117)$$

There are two cases:

(i) Relation (2.104) holds for each integer $q \in \{0, \ldots, Q - 2\}$;
(ii) There is an integer $q \in \{0, \ldots, Q - 2\}$ which does not satisfy (2.104).

Consider the case (i). By the choice of $\{\tau_i\}_{i=0}^Q$, (2.116), which holds with $q = Q - 1$, (2.117), (2.107), (2.100), and (2.94),

$$Q - 1 = \tau_Q - T_1,$$
$$I^g(T_1, T_2, x, u) - I^g(T_1, T_2, x_*, u_*)$$
$$\geq (3/4)(Q - 1)D_1 - D_3 - (c_1 + 2)(\Lambda_0 + a_0)$$
$$\geq 2^{-1}\Delta D_1 - D_3 - (c_1 + 2)(\Lambda_0 + a_0) \geq 2S$$

and (2.93) holds.

Consider the case (ii). It is easy to see that (2.108) holds. Relation (2.93) follows from (2.116), which holds with $q = Q - 1$, (2.108), (2.107), and (2.97). This completes the proof of Lemma 2.17.   □

Lemma 2.18. *There exists $\gamma > 0$ such that for each $\delta \in (0, \gamma)$ and each $S > 0$ there exist $\Delta \geq 1$ and a neighborhood $\mathcal{U}$ of f_r in $\mathfrak{M}$ such that for each $g \in \mathcal{U}$, each $T_1 \in R^1$, each $T_2 \geq T_1 + \Delta$, and each trajectory-control pair*

$$x : [T_1, T_2] \to R^n, \; u : [T_1, T_2] \to R^m$$

which satisfies

$$|x(T_i) - x_*(T_i)| \leq \gamma, \; i = 1, 2,$$
$$|x(t) - x_*(t)| > \delta \text{ for all } t \in [T_1, T_2] \tag{2.118}$$

the inequality

$$I^g(T_1, T_2, x, u) > U^g(T_1, T_2, x(T_1), x(T_2)) + S$$

holds.

Proof. There exists

$$\gamma \in (0, \min\{8^{-1}, d_0\})$$

such that property (iv) (see the definition of f) holds with $\epsilon = 1$ and $\delta = \gamma$.

Let $\delta \in (0, \gamma)$ and $S > 0$. By Proposition 2.9 there exists a neighborhood $\mathcal{U}_1$ of f_r in $\mathfrak{M}$ such that for each $g \in \mathcal{U}_1$, each $T_1 \in R^1$, each $T_2 \in [T_1 + 8^{-1}, T_1 + 8(b_* + 2)]$, and each trajectory-control pair

$$x : [T_1, T_2] \to R^n, \; u : [T_1, T_2] \to R^m$$

which satisfies

$$\min\{I^{f_r}(T_1, T_2, x, u), \; I^g(T_1, T_2, x, u)\} \leq 16(\Lambda_0 + a_0 + 4)(2b_* + 6) \tag{2.119}$$

the inequality

$$|I^g(T_1, T_2, x, u) - I^{f_r}(T_1, T_2, x, u)| \leq 1 \tag{2.120}$$

holds.

Fix a number

$$S_1 > 2S + 1 + 2[1 + \Lambda_0(4 + b_*) + 4a_0 + b_* + 1] + 2a_0(1 + b_*). \tag{2.121}$$

By Lemma 2.17 there exist $\Delta_1 \geq 1$ and a neighborhood $\mathcal{U}$ of f_r in $\mathfrak{M}$ such that $\mathcal{U} \subset \mathcal{U}_1$ and for each $g \in \mathcal{U}$, each $T_1 \in R^1$, each $T_2 \geq T_1 + \Delta_1$, and each trajectory-control pair

$$x : [T_1, T_2] \to R^n, \ u : [T_1, T_2] \to R^m$$

which satisfies

$$|x(t) - x_*(t)| \geq \delta \text{ for all } t \in [T_1, T_2] \tag{2.122}$$

the following inequality holds:

$$I^g(T_1, T_2, x, u) > I^g(T_1, T_2, x_*, u_*) + S_1. \tag{2.123}$$

Fix a number

$$\Delta > \Delta_1 + 4(4 + 4b_*) + 16. \tag{2.124}$$

Assume that $g \in \mathcal{U}$, $T_1 \in R^1$, each $T_2 \geq T_1 + \Delta$ and a trajectory-control pair

$$x : [T_1, T_2] \to R^n, \ u : [T_1, T_2] \to R^m$$

satisfies (2.118). We show that

$$I^g(T_1, T_2, x, u) > U^g(T_1, T_2, x(T_1), x(T_2)) + S. \tag{2.125}$$

By (2.118), the choice of γ and property (iv) there exist

$$\tau_1 \in (T_1, T_1 + b_*] \text{ and } \tau_2 \in [T_2 - b_*, T_2) \tag{2.126}$$

and trajectory-control pairs

$$x_1 : [T_1, \tau_1] \to R^n, \ u_1 : [T_1, \tau_1] \to R^m,$$
$$x_2 : [\tau_2, T_2] \to R^n, \ u_2 : [\tau_2, T_2] \to R^m$$

such that

$$x_i(T_i) = x(T_i), \ i = 1, 2,$$
$$x_i(\tau_i) = x_*(\tau_i), \ i = 1, 2,$$
$$|x_1(t) - x_*(t)| \leq 1 \text{ for all } t \in [T_1, \tau_1],$$
$$|x_2(t) - x_*(t)| \leq 1 \text{ for all } t \in [\tau_2, T_2],$$
$$I^f(T_1, \tau_1, x_1, u_1) \leq I^f(T_1, \tau_1, x_*, u_*) + 1,$$
$$I^f(\tau_2, T_2, x_2, u_2) \leq I^f(\tau_2, T_2, x_*, u_*) + 1. \tag{2.127}$$

Define a trajectory-control pair

$$x_3 : [T_1, T_2] \to R^n, \; u_3 : [T_1, T_2] \to R^m$$

by

$$x_3(t) = x_1(t), \; u_3(t) = u_1(t) \text{ for all } t \in [T_1, \tau_1],$$
$$x_3(t) = x_*(t), \; u_3(t) = u_*(t) \text{ for all } t \in (\tau_1, \tau_2],$$
$$x_3(t) = x_2(t), \; u_3(t) = u_2(t) \text{ for all } t \in (\tau_2, T_2]. \tag{2.128}$$

Relations (2.126)–(2.128) imply that

$$U^g(T_1, T_2, x(T_1), x(T_2)) \le I^g(T_1, T_2, x_3, u_3). \tag{2.129}$$

By (2.124), (2.127), (2.128), (2.126), (2.38), and (2.39),

$$I^{f_r}(T_1, T_1 + 1 + b_*, x_3, u_3) \le I^f(T_1, T_1 + 1 + b_*, x_3, u_3) + b_*$$
$$\le I^f(T_1, T_1 + b_* + 1, x_*, u_*) + b_* + 1$$

and

$$I^{f_r}(T_2 - 1 - b_*, T_2, x_3, u_3) \le I^f(T_2 - 1 - b_*, T_2, x_3, u_3) + b_*$$
$$\le I^f(T_2 - 1 - b_*, T_2, x_*, u_*) + b_* + 1.$$

By these relations, (2.40) and the choice of $\mathcal{U}_1$ [see (2.119), (2.120)]

$$\max\{I^g(T_1, T_1 + 1 + b_*, x_3, u_3), \; I^g(T_2 - 1 - b_*, T_2, x_3, u_3)\}$$
$$\le 1 + \Lambda_0(4 + b_*) + 4a_0 + b_* + 1. \tag{2.130}$$

It follows from (2.118), (2.124), and the choice of $\mathcal{U}$ and Δ_1 [see (2.122), (2.123)] that

$$I^g(T_1 + 1 + b_*, T_2 - 1 - b_*, x, u) > I^g(T_1 + 1 + b_*, T_2 - 1 - b_*, x_*, u_*) + S_1.$$

It follows from this relation, (2.128), (2.126), (2.130), (2.121), and assumption (A) that

$$I^g(T_1, T_2, x, u) - I^g(T_1, T_2, x_3, u_3)$$
$$> S_1 - 2[1 + \Lambda_0(4 + b_*) + 4a_0 + b_* + 1] - 2a_0(1 + b_*) > 2S + 1.$$

Together with (2.129) this implies (2.125). This completes the proof of Lemma 2.18.
□

Lemma 2.19. *Let $\epsilon \in (0, \min\{1, d_0\})$. Then there exist $\delta \in (0, \epsilon)$, $\Delta \geq 1$ and a neighborhood $\mathcal{U}$ of f_r in $\mathfrak{M}$ such that for each $g \in \mathcal{U}$, each $T_1 \in R^1$, each $T_2 \geq T_1 + \Delta$, and each trajectory-control pair*

$$x : [T_1, T_2] \to R^n, \ u : [T_1, T_2] \to R^m$$

which satisfies

$$|x(T_i) - x_*(T_i)| \leq \delta, \ i = 1, 2,$$
$$I^g(T_1, T_2, x, u) \leq U^g(T_1, T_2, x(T_1), x(T_2)) + \delta \qquad (2.131)$$

the following inequality holds:

$$|x(t) - x_*(t)| \leq \epsilon \text{ for all } t \in [T_1, T_2].$$

Proof. There exist $\Delta_1 \geq 1$ and $\delta_0 \in (0, \epsilon)$ such that Lemma 2.16 holds with $\delta = \delta_0$. Let a number $\gamma > 0$ be as guaranteed in Lemma 2.18. Choose a number

$$\delta \in (0, 8^{-1} \min\{\delta_0, \gamma, 1\}). \qquad (2.132)$$

By Lemma 2.18 there exist $\Delta_2 \geq \Delta_1 + 1$ and a neighborhood $\mathcal{U}_1$ of f_r in $\mathfrak{M}$ such that for each $g \in \mathcal{U}_1$, each $T_1 \in R^1$, each $T_2 \geq T_1 + \Delta_2$, and each trajectory-control pair

$$x : [T_1, T_2] \to R^n, \ u : [T_1, T_2] \to R^m$$

which satisfies

$$|x(T_i) - x_*(T_i)| \leq \gamma, \ i = 1, 2,$$
$$|x(t) - x_*(t)| > \delta \text{ for all } t \in [T_1, T_2] \qquad (2.133)$$

the inequality

$$I^g(T_1, T_2, x, u) > U^g(T_1, T_2, x(T_1), x(T_2)) + 1 \qquad (2.134)$$

holds.

Since Lemma 2.16 holds with $\delta = \delta_0$ and Δ_1 there exists a neighborhood $\mathcal{U}$ of f_r in $\mathfrak{M}$ such that

$$\mathcal{U} \subset \mathcal{U}_1$$

and for each $g \in \mathcal{U}$, each $T_1 \in R^1$, each

$$T_2 \in [T_1 + \Delta_1, T_1 + 2\Delta_1 + 2\Delta_2 + 8]$$

and each trajectory-control pair

$$x : [T_1, T_2] \to R^n, \ u : [T_1, T_2] \to R^m$$

which satisfies

$$|x(T_i) - x_*(T_i)| \leq \delta_0, \ i = 1, 2,$$
$$I^g(T_1, T_2, x, u) \leq U^g(T_1, T_2, x(T_1), x(T_2)) + \delta_0 \tag{2.135}$$

the following inequality holds:

$$|x(t) - x_*(t)| \leq \epsilon \text{ for all } t \in [T_1, T_2]. \tag{2.136}$$

Choose a number

$$\Delta > 8\Delta_2 + 8\Delta_1 + 8. \tag{2.137}$$

Assume that

$$g \in \mathcal{U}, \ T_1 \in R^1, \ T_2 \geq T_1 + \Delta$$

and that a trajectory-control pair

$$x : [T_1, T_2] \to R^n, \ u : [T_1, T_2] \to R^m$$

satisfies (2.131). Let

$$t_1 \in [T_1, T_2 - \Delta_1 - \Delta_2 - 1]. \tag{2.138}$$

We show that there exists a number t_2 such that

$$t_2 \in [t_1 + \Delta_1, t_1 + \Delta_1 + \Delta_2],$$
$$|x(t_2) - x_*(t_2)| \leq \delta. \tag{2.139}$$

Assume the contrary and set

$$\tilde{t}_1 = \sup\{t \in [T_1, t_1 + \Delta_1] : \ |x(t) - x_*(t)| \leq \delta\},$$
$$\tilde{t}_2 = \inf\{t \in [t_1 + \Delta_1, T_2] : \ |x(t) - x_*(t)| \leq \delta\}. \tag{2.140}$$

Then

$$\tilde{t}_1 < t_1 + \Delta_1, \ \tilde{t}_2 > t_1 + \Delta_1 + \Delta_2,$$
$$|x(\tilde{t}_i) - x_*(\tilde{t}_i)| = \delta, \ i = 1, 2.$$

There exist

$$b_1 \in (\tilde{t}_1, t_1 + \Delta_1) \text{ and } b_2 \in (t_1 + \Delta_1 + \Delta_2, \tilde{t}_2) \qquad (2.141)$$

such that

$$|x(b_i) - x_*(b_i)| < \gamma, \ i = 1, 2. \qquad (2.142)$$

It is easy to see that

$$|x(t) - x_*(t)| > \delta \text{ for all } t \in [b_1, b_2]. \qquad (2.143)$$

By (2.141)–(2.143) and the choice of $\mathcal{U}_1$, Δ_2 [see (2.133), (2.134)],

$$I^g(b_1, b_2, x, u) > U^g(b_1, b_2, x(b_1), x(b_2)) + 1.$$

This contradicts (2.131). The contradiction we have reached proves that for each number t_1 satisfying (2.138) there exists a number t_2 which satisfies (2.139). Therefore there exists a sequence of numbers $\{t_j\}_{j=1}^{Q}$ such that

$$t_1 = T_1, \ t_Q = T_2,$$
$$t_{j+1} - t_j \in [\Delta_1, 2\Delta_1 + 2\Delta_2 + 4], \ j = 1, \ldots, Q - 1,$$
$$|x(t_j) - x_*(t_j)| \le \delta, \ j = 1, \ldots, Q. \qquad (2.144)$$

In view of (2.131),

$$I^g(t_j, t_{j+1}, x, u) \le U^g(t_j, t_{j+1}, x(t_j), x(t_{j+1})) + 2\delta.$$

It follows from this inequality, (2.132), (2.144) and the choice of $\mathcal{U}$ [see (2.133), (2.136)] that for all $j = 1, \ldots, Q - 1$,

$$|x(t) - x_*(t)| \le \epsilon \text{ for all } t \in [t_j, t_{j+1}].$$

This completes the proof of Lemma 2.19.   □

Lemma 2.20. *Let $S > 0$ and $\epsilon \in (0, 1)$. Then there exist $\Delta \ge 1$, $\delta \in (0, \epsilon)$ and a neighborhood $\mathcal{U}$ of f_r in $\mathfrak{M}$ such that for each $g \in \mathcal{U}$, each $T_1 \in R^1$, each $T_2 \ge T_1 + 2\Delta$, and each trajectory-control pair*

$$x : [T_1, T_2] \to R^n, \ u : [T_1, T_2] \to R^m$$

which satisfies

$$I^g(T_1, T_2, x, u) \leq \inf\{\sigma^g(T_1, T_2, y) : (T_1, y) \in A\} + S,$$
$$I^g(T_1, T_2, x, u) \leq U^g(T_1, T_2, x(T_1), x(T_2)) + \delta \qquad (2.145)$$

the following inequality holds:

$$|x(t) - x_*(t)| \leq \epsilon \text{ for all } t \in [T_1 + \Delta, T_2 - \Delta]. \qquad (2.146)$$

Moreover, if

$$|x(T_1) - x_*(T_1)| \leq \delta,$$

then

$$|x(t) - x_*(t)| \leq \epsilon \text{ for all } t \in [T_1, T_2 - \Delta]$$

and if

$$|x(T_2) - x_*(T_2)| \leq \delta,$$

then

$$|x(t) - x_*(t)| \leq \epsilon \text{ for all } t \in [T_1 + \Delta, T_2].$$

Proof. By property (ii) (see the definition of f), assumption (A) and Proposition 2.9, there exists a neighborhood $\mathcal{U}_0$ of f_r in $\mathfrak{M}$ such that for each $g \in \mathcal{U}_0$, each $T_1 \in R^1$, each $T_2 \geq T_1 + 8^{-1}$,

$$\inf\{\sigma^g(T_1, T_2, y) : (T_1, y) \in A\} < \infty. \qquad (2.147)$$

There exists $\delta_1 > 0$ such that property (iv) (see the definition of f) holds with $\epsilon = 1$, $\delta = \delta_1$. By Proposition 2.9 there exists a neighborhood $\mathcal{U}_1$ of f_r in $\mathfrak{M}$ such that

$$\mathcal{U}_1 \subset \mathcal{U}_0$$

and for each $g \in \mathcal{U}_1$, each $T_1 \in R^1$, each $T_2 \in [T_1 + 4^{-1}b_*, T_1 + 4b_*]$, and each trajectory-control pair

$$x : [T_1, T_2] \to R^n, \ u : [T_1, T_2] \to R^m$$

which satisfies

$$\min\{I^{f_r}(T_1, T_2, x, u), \ I^g(T_1, T_2, x, u)\}$$
$$\leq 4 + 4b_* + 4\Lambda_0(b_* + 1) + 4a_0 \tag{2.148}$$

the inequality

$$|I^{f_r}(T_1, T_2, x, u) - I^g(T_1, T_2, x, u)| \leq 1 \tag{2.149}$$

holds.

By Lemma 2.19 there exist

$$\delta \in (0, \min\{1, \ 2^{-1}\delta_1, \ d_0, \ \epsilon\}), \ \Delta_0 \geq 1 \tag{2.150}$$

and a neighborhood $\mathcal{U}_2$ of f_r in $\mathfrak{M}$ such that

$$\mathcal{U}_2 \subset \mathcal{U}_1$$

and for each $g \in \mathcal{U}_2$, each $T_1 \in R^1$, each $T_2 \geq T_1 + \Delta_0$ and each trajectory-control pair

$$x : [T_1, T_2] \to R^n, \ u : [T_1, T_2] \to R^m$$

which satisfies

$$|x(T_i) - x_*(T_i)| \leq \delta, \ i = 1, 2,$$
$$I^g(T_1, T_2, x, u) \leq U^g(T_1, T_2, x(T_1), x(T_2)) + \delta \tag{2.151}$$

the following inequality holds:

$$|x(t) - x_*(t)| \leq \epsilon \text{ for all } t \in [T_1, T_2]. \tag{2.152}$$

By Lemma 2.17, there exist $\Delta_1 \geq 1$ and a neighborhood $\mathcal{U}$ of f_r in $\mathfrak{M}$ such that

$$\mathcal{U} \subset \mathcal{U}_2$$

and for each $g \in \mathcal{U}$, each $T_1 \in R^1$, each $T_2 \geq T_1 + \Delta_1$, and each trajectory-control pair

$$x : [T_1, T_2] \to R^n, \ u : [T_1, T_2] \to R^m$$

which satisfies

$$|x(t) - x_*(t)| \geq \delta \text{ for all } t \in [T_1, T_2] \tag{2.153}$$

the following inequality holds:

$$I^g(T_1, T_2, x, u) > I^g(T_1, T_2, x_*, u_*)$$
$$+ S + 10 + (b_* + 4)(4 + 4\Lambda_0 + 4a_0) + 4a_0 + 4.$$

Fix a number

$$\Delta > 128(b_* + 1 + \Delta_0 + \Delta_1). \tag{2.154}$$

Assume that

$$g \in \mathcal{U}, \ T_1 \in R^1, \ T_2 \geq T_1 + 2\Delta$$

and that a trajectory-control pair

$$x : [T_1, T_2] \to R^n, \ u : [T_1, T_2] \to R^m$$

satisfies (2.145).

By the definition of Δ_1, $\mathcal{U}$ [see (2.153)], (2.145) and (2.154) there exists $\tilde{t} \in [T_1, T_2]$ which satisfies

$$|x(\tilde{t}) - x_*(\tilde{t})| \leq \delta.$$

Set

$$t_1 = \inf\{t \in [T_1, T_2] : \ |x(t) - x_*(t)| \leq \delta\},$$
$$t_2 = \sup\{t \in [T_1, T_2] : \ |x(t) - x_*(t)| \leq \delta\}. \tag{2.155}$$

It follows from (2.155), property (iv), and the choice of δ_1 that there exist

$$\tau_2 \in (t_2, t_2 + b_*] \text{ and } \tau_1 \in [t_1 - b_*, t_1), \tag{2.156}$$

and trajectory-control pairs

$$x_1 : [\tau_1, t_1] \to R^n, \ u_1 : [\tau_1, t_1] \to R^m,$$
$$x_2 : [t_2, \tau_2] \to R^n, \ u_2 : [t_2, \tau_2] \to R^m$$

such that

$$x_i(t_i) = x(t_i), \ i = 1, 2,$$
$$x_i(\tau_i) = x_*(\tau_i), \ i = 1, 2,$$
$$|x_1(t) - x_*(t)| \leq 1 \text{ for all } t \in [\tau_1, t_1],$$

$$|x_2(t) - x_*(t)| \leq 1 \text{ for all } t \in [t_2, \tau_2],$$

$$I^f(\tau_1, t_1, x_1, u_1) \leq I^f(\tau_1, t_1, x_*, u_*) + 1,$$

$$I^f(t_2, \tau_2, x_2, u_2) \leq I^f(t_2, \tau_2, x_*, u_*) + 1. \tag{2.157}$$

We may assume without loss of generality that

$$\tau_2 = t_2 + b_*, \ \tau_1 = t_1 - b_*. \tag{2.158}$$

By (2.158), (2.157), (2.38), and (2.39),

$$I^{fr}(\tau_1, t_1, x_1, u_1) \leq I^f(\tau_1, t_1, x_1, u_1) + b_* \leq I^f(\tau_1, t_1, x_*, u_*) + b_* + 1,$$

$$I^{fr}(t_2, \tau_2, x_2, u_2) \leq I^f(t_2, \tau_2, x_2, u_2) + b_* \leq I^f(t_2, \tau_2, x_*, u_*) + b_* + 1. \tag{2.159}$$

By these relations, (2.158) and the choice of $\mathcal{U}_1$ [see (2.148), (2.149)],

$$|I^g(\tau_1, t_1, y, v) - I^{fr}(\tau_1, t_1, y, v)| \leq 1, \ (y, v) \in \{(x_1, u_1), \ (x_*, u_*)\},$$

$$|I^g(t_2, \tau_2, y, v) - I^{fr}(t_2, \tau_2, y, v)| \leq 1, \ (y, v) \in \{(x_2, u_2), \ (x_*, u_*)\}.$$

Together with (2.159) these relations imply that

$$I^g(\tau_1, t_1, x_1, u_1) \leq I^g(\tau_1, t_1, x_*, u_*) + b_* + 3,$$

$$I^g(t_2, \tau_2, x_2, u_2) \leq I^g(t_2, \tau_2, x_*, u_*) + b_* + 3. \tag{2.160}$$

We show that

$$t_1 - T_1 \leq 8^{-1}\Delta.$$

Assume the contrary. Then

$$t_1 - T_1 > 8^{-1}\Delta. \tag{2.161}$$

Define a trajectory-control pair

$$x_3 : [T_1, T_2] \to R^n, \ u_3 : [T_1, T_2] \to R^m$$

by

$$x_3(t) = x_*(t), \ u_3(t) = u_*(t) \text{ for all } t \in [T_1, \tau_1],$$

$$x_3(t) = x_1(t), \ u_3(t) = u_1(t) \text{ for all } t \in (\tau_1, t_1],$$

$$x_3(t) = x(t), \ u_3(t) = u(t) \text{ for all } t \in (t_1, T_2]. \tag{2.162}$$

Relations (2.161), (2.158), (2.154) and (2.155) imply that

$$\tau_1 - T_1 \geq 2\Delta_0 + 2\Delta_1 + 2,$$
$$|x(t) - x_*(t)| \geq \delta \text{ for all } t \in [T_1, \tau_1].$$

By these relations and the choice of $\mathcal{U}$ and Δ_1 [see (2.153)]

$$I^g(T_1, \tau_1, x, u) > I^g(T_1, \tau_1, x_*, u_*)$$
$$+S + 10 + (b_* + 4)(4 + 4\Lambda_0 + 4a_0) + 4a_0 + 4.$$

It follows from this inequality, (2.162), (2.160), (2.158), and assumption (A) that

$$I^g(T_1, T_2, x, u) - I^g(T_1, T_2, x_3, u_3) > S + 10.$$

This contradicts (2.145). The contradiction we have reached proves that

$$t_1 - T_1 \leq 8^{-1}\Delta. \tag{2.163}$$

Analogously we can show that

$$T_2 - t_2 \leq 8^{-1}\Delta. \tag{2.164}$$

It follows from the choice of $\mathcal{U}_2$ and δ [see (2.150)–(2.152)], (2.163), (2.164), (2.154), (2.145), and (2.155) that

$$|x(t) - x_*(t)| \leq \epsilon \text{ for all } t \in [t_1, t_2].$$

This completes the proof of Lemma 2.20.   □

Proposition 2.9, property (ii) (see the definition of f), and assumption (A) imply the following result.

Lemma 2.21. *There exists a neighborhood $\mathcal{U}$ of f_r in $\mathfrak{M}$ such that for each $g \in \mathcal{U}$, each $T_1 \in R^1$, each $T_2 > T_1$,*

$$\inf\{U^g(T_1, T_2, y_1, y_2) : (T_i, y_i) \in A, \ i = 1, 2\} \leq I^g(T_1, T_2, x_*, u_*) < \infty. \tag{2.165}$$

Lemma 2.22. *Let $S > 0$. Then there exist a neighborhood $\mathcal{U}$ of f_r in $\mathfrak{M}$ and numbers $\delta, Q > 0$ such that for each $g \in \mathcal{U}$, each $T_1 \in R^1$, each $T_2 \geq T_1 + 1$ and each trajectory-control pair*

$$x : [T_1, T_2] \to R^n, \ u : [T_1, T_2] \to R^m$$

which satisfies

$$I^g(T_1, T_2, x, u) \leq \inf\{U^g(T_1, T_2, y_1, y_2) : (T_i, y_i) \in A, \ i = 1, 2\} + S,$$
$$I^g(T_1, T_2, x, u) \leq U^g(T_1, T_2, x(T_1), x(T_2)) + \delta \qquad (2.166)$$

the following inequality holds:

$$|x(t)| \leq Q \text{ for all } t \in [T_1, T_2]. \qquad (2.167)$$

Proof. By Lemma 2.21 there exists a neighborhood $\mathcal{U}_0$ of f_r in $\mathfrak{M}$ such that (2.165) holds for each $g \in \mathcal{U}_0$, each $T_1 \in R^1$, and each $T_2 > T_1$. There exists a number

$$\delta_0 \in (0, 8^{-1} \min\{1, d_0\}) \qquad (2.168)$$

such that property (iv) (see the definition of f) holds with $\epsilon = 1$, $\delta = \delta_0$. Fix

$$S_1 > S + 1. \qquad (2.169)$$

There exist

$$\Delta_0 \geq 16 + 8b_*, \ \delta_1 \in (0, \delta_0)$$

and a neighborhood $\mathcal{U}_1$ of f_r in $\mathcal{M}$ such that Lemma 2.20 holds with

$$S = 4S_1 + 8, \ \epsilon = \delta_0, \ \Delta = \Delta_0,$$
$$\delta = \delta_1, \ \mathcal{U} = \mathcal{U}_1. \qquad (2.170)$$

By Proposition 2.9 there exists a neighborhood $\mathcal{U}$ of f_r in $\mathfrak{M}$ such that

$$\mathcal{U} \subset \mathcal{U}_1 \cap \mathcal{U}_2$$

and for each $g \in \mathcal{U}$, each $T_1 \in R^1$, each

$$T_2 \in [T_1 + 8^{-1} \min\{1, b_*\}, \ T_1 + 2\Delta_0 + 8]$$

and each trajectory-control pair

$$x : [T_1, T_2] \to R^n, \ u : [T_1, T_2] \to R^m$$

which satisfies

$$\min\{I^g(T_1, T_2, x, u), \ I^{f_r}(T_1, T_2, x, u)\}$$
$$\leq 4 + 4b_* + (4\Delta_0 + 4 + 4a_0)(3\Delta_0 + 4 + 4b_*) \qquad (2.171)$$

the inequality

$$|I^g(T_1, T_2, x, u) - I^{f_r}(T_1, T_2, x, u)| \leq 1 \qquad (2.172)$$

holds.

By Proposition 2.7 there exists

$$D_1 > 100 + \sup\{|x_*(t)| : t \in R^1\} \qquad (2.173)$$

such that for each $g \in \mathfrak{M}$, each $T_1 \in R^1$, each

$$T_2 \in [T_1 + 8^{-1}, T_1 + 100(\Delta_0 + 1)]$$

and each trajectory-control pair

$$x : [T_1, T_2] \to R^n, \ u : [T_1, T_2] \to R^m$$

which satisfies

$$I^g(T_1, T_2, x, u) \leq S + 2b_* + (\Lambda_0 + 4)(8\Delta_0 + 8) + 4a_0 \qquad (2.174)$$

the following relation holds:

$$|x(t)| \leq D_1 \text{ for all } t \in [T_1, T_2]. \qquad (2.175)$$

Choose numbers

$$\delta \in (0, 4^{-1}\delta_1), \ Q > 4D_1 + 4. \qquad (2.176)$$

Assume that

$$g \in \mathcal{U}, \ T_1 \in R^1, \ T_2 \geq T_1 + 1$$

and that a trajectory-control pair

$$x : [T_1, T_2] \to R^n, \ u : [T_1, T_2] \to R^m$$

satisfies (2.166). We show that (2.167) holds. There are two cases:

$$(1) \ T_2 - T_1 \geq 4\Delta_0 + 4;$$
$$(2) \ T_2 - T_1 < 4\Delta_0 + 4.$$

Consider the case (1). It follows from Lemma 2.20 and the definition of $\Delta_0 \geq 1$, $\delta_1 \in (0, \delta_0)$ [see (2.170)], (2.166), (2.176), and (2.169) that

$$|x(t) - x_*(t)| \leq \delta_0 \text{ for all } t \in [T_1 + \Delta_0, T_2 - \Delta_0]. \qquad (2.177)$$

By property (iv) and the choice of δ_0 there exist

$$\tau_1 \in [T_1 + \Delta_0 - b_*, T_1 + \Delta_0) \text{ and } \tau_2 \in (T_2 - \Delta_0, T_2 - \Delta_0 + b_*] \qquad (2.178)$$

and trajectory-control pairs

$$x_1 : [\tau_1, T_1 + \Delta_0] \to R^n, \ u_1 : [\tau_1, T_1 + \Delta_0] \to R^m,$$
$$x_2 : [T_2 - \Delta_0, \tau_2] \to R^n, \ u_2 : [T_2 - \Delta_0, \tau_2] \to R^m$$

such that

$$x_i(\tau_i) = x_*(\tau_i), \ i = 1, 2,$$
$$x_1(T_1 + \Delta_0) = x(T_1 + \Delta_0),$$
$$x_2(T_2 - \Delta_0) = x(T_2 - \Delta_0),$$
$$I^f(\tau_1, T_1 + \Delta_0, x_1, u_1) \leq I^f(\tau_1, T_1 + \Delta_0, x_*, u_*) + 1,$$
$$I^f(T_2 - \Delta_0, \tau_2, x_2, u_2) \leq I^f(T_2 - \Delta_0, \tau_2, x_*, u_*) + 1,$$
$$|x_1(t) - x_*(t)| \leq 1 \text{ for all } t \in [\tau_1, T_1 + \Delta_0],$$
$$|x_2(t) - x_*(t)| \leq 1 \text{ for all } t \in [T_2 - \Delta_0, \tau_2]. \qquad (2.179)$$

We may assume that

$$\tau_1 = T_1 + \Delta_0 - b_*, \ \tau_2 = T_2 - \Delta_0 + b_*. \qquad (2.180)$$

Define a trajectory-control pair

$$x_3 : [T_1, T_2] \to R^n, \ u_3 : [T_1, T_2] \to R^m$$

and

$$x_4 : [T_1, T_2] \to R^n, \ u_4 : [T_1, T_2] \to R^m$$

by

$$x_3(t) = x_*(t), \ u_3(t) = u_*(t) \text{ for all } t \in [T_1, \tau_1],$$
$$x_3(t) = x_1(t), \ u_3(t) = u_1(t) \text{ for all } t \in (\tau_1, T_1 + \Delta_0],$$
$$x_3(t) = x(t), \ u_3(t) = u(t) \text{ for all } t \in (T_1 + \Delta_0, T_2],$$
$$x_4(t) = x(t), \ u_4(t) = u(t) \text{ for all } t \in [T_1, T_2 - \Delta_0],$$
$$x_4(t) = x_2(t), \ u_4(t) = u_2(t) \text{ for all } t \in (T_2 - \Delta_0, \tau_2],$$
$$x_4(t) = x_*(t), \ u_4(t) = u_*(t) \text{ for all } t \in (\tau_2, T_2]. \qquad (2.181)$$

Relations (2.180), (2.179), and (2.181) imply that

$$I^{fr}(T_1, T_1 + \Delta_0, x_3, u_3) \le I^f(T_1, T_1 + \Delta_0, x_3, u_3) + b_*$$
$$\le I^f(T_1, T_1 + \Delta_0, x_*, u_*) + b_* + 1,$$
$$I^{fr}(T_2 - \Delta_0, T_2, x_4, u_4) \le I^f(T_2 - \Delta_0,, T_2, x_4, u_4) + b_*$$
$$\le I^f(T_2 - \Delta_0, T_2, x_*, u_*) + b_* + 1. \tag{2.182}$$

By these relations, the choice of $\mathcal{U}$ [see (2.171), (2.172)], (2.180),

$$|I^{fr}(T_1, T_1 + \Delta_0, y, v) - I^g(T_1, T_1 + \Delta_0, y, v)| \le 1,$$
$$(y, v) \in \{(x_3, u_3), (x_*, u_*)\},$$
$$|I^{fr}(T_2 - \Delta_0, T_2, y, v) - I^g(T_2 - \Delta_0, T_2, y, v)| \le 1,$$
$$(y, v) \in \{(x_4, u_4), (x_*, u_*)\}.$$

Together with (2.182) this implies that

$$I^g(T_1, T_1 + \Delta_0, x_3, u_3) \le I^f(T_1, T_1 + \Delta_0, x_*, u_*) + b_* + 3,$$
$$I^g(T_2 - \Delta_0, T_2, x_4, u_4) \le I^f(T_2 - \Delta_0, T_2, x_*, u_*) + b_* + 3. \tag{2.183}$$

It follows from (2.166), (2.181) and (2.183) that

$$S \ge I^g(T_1, T_2, x, u) - I^g(T_1, T_2, x_3, u_3)$$
$$= I^g(T_1, T_1 + \Delta_0, x, u) - I^g(T_1, T_1 + \Delta_0, x_3, u_3)$$
$$\ge I^g(T_1, T_1 + \Delta_0, x, u) - b_* - 3 - \Lambda_0(\Delta_0 + 2) - 2a_0$$

and

$$I^g(T_1, T_1 + \Delta_0, x, u) \le S + b_* + 3 + \Lambda_0(\Lambda_0 + 2) + 2a_0. \tag{2.184}$$

Analogously,

$$I^g(T_2 - \Delta_0, T_2, x, u) \le S + b_* + 3 + \Lambda_0(\Delta_0 + 2) + 2a_0. \tag{2.185}$$

By (2.184), (2.185) and the choice of D_1 [see (2.173)–(2.175)]

$$|x(t)| \le D_1 \text{ for all } t \in [T_1, T_1 + \Delta_0] \cup [T_2 - \Delta_0, T_2].$$

Together with (2.177), (2.173), (2.168), and (2.176) this implies (2.167).

Consider the case (2). By (2.166), the choice of $\mathcal{U}$ [see (2.171), (2.172)], and assumption (A),

$$I^g(i, i+1, x_*, u_*) \leq \Lambda_0 + 1, \ i = 0, 1, \ldots$$

and

$$I^g(T_1, T_2, x, u) \leq I^g(T_1, T_2, x_*, u_*) + S$$
$$\leq S + 2a_0 + (\Lambda_0 + 1)(4\Lambda_0 + 6).$$

Relation (2.167) follows from the relation above, the choice of D_1, (2.175) and (2.176). This completes the proof of Lemma 2.22. $\square$

Lemma 2.23. *Let $\epsilon_0, S > 0$. Then there exist $\Delta > 0$ and a neighborhood $\mathcal{U}$ of f_r in $\mathfrak{M}$ such that for each $g \in \mathcal{U}$, each $T_1 \in R^1$, each $T_2 \geq T_1 + \Delta$, and each trajectory-control pair*

$$x : [T_1, T_2] \to R^n, \ u : [T_1, T_2] \to R^m$$

which satisfies

$$I^g(T_1, T_2, x, u) \leq \inf\{\sigma^g(T_1, T_2, y) : \ (T_1, y) \in A\} + S, \tag{2.186}$$

each $\tau \in [T_1, T_2 - \Delta]$,

$$\min\{|x(t) - x_*(t)| : \ t \in (\tau, \tau + \Delta)\} < \epsilon_0. \tag{2.187}$$

Proof. We may assume without loss of generality that

$$\epsilon_0 < 2^{-1} \min\{1, d_0, b_*\}. \tag{2.188}$$

We may also assume without loss of generality that property (iv) holds with $\epsilon = 1$, $\delta = \epsilon_0$.

By property (ii) and assumption (A) there is $\tilde{\Lambda} > 0$ such that

$$I^f(T, T+b, x_*, u_*) \leq \tilde{\Lambda} \text{ for all } T \in R^1$$

$$\text{and all } b \in [2^{-1} \min\{b_*, 1\}, \ 2b_* + 2]. \tag{2.189}$$

By Proposition 2.9 there exists a neighborhood $\mathcal{U}_1$ of f_r in $\mathfrak{M}$ such that for each $g \in \mathcal{U}_1$, each pair of numbers

$$S_1 \in R^1, \ S_2 \in [S_1 + 2^{-1} \min\{b_*, 1\}, \ S_1 + 2b_* + 2]$$

and each trajectory-control pair

$$x : [S_1, S_2] \to R^n, \; u : [S_1, S_2] \to R^m$$

which satisfies

$$\min\{I^g(S_1, S_2, x, u), \; I^{f_r}(S_1, S_2, x, u)\}$$

$$\le \tilde{2}\Lambda + 2 + 2b_* \tag{2.190}$$

the inequality

$$|I^g(S_1, S_2, x, u) - I^{f_r}(S_1, S_2, x, u)| \le 1 \tag{2.191}$$

holds.

By (2.188) and Lemma 2.17 there exist $\Delta_1 \ge 1$ and a neighborhood $\mathcal{U}$ of f_r in $\mathfrak{M}$ such that

$$\mathcal{U} \subset \mathcal{U}_1$$

and for each $g \in \mathcal{U}$, each $S_1 \in R^1$, each $S_2 \ge S_1 + \Delta_1$, and each trajectory-control pair

$$y : [S_1, S_2] \to R^n, \; v : [S_1, S_2] \to R^m$$

which satisfies

$$|y(t) - x_*(t)| \ge \epsilon_0 \text{ for all } t \in [S_1, S_2] \tag{2.192}$$

the following inequality holds:

$$I^g(S_1, S_2, y, v) > I^g(S_1, S_2, x_*, u_*)$$

$$+ S + 2\tilde{\Lambda} + 8 + 2b_*(1 + a_0). \tag{2.193}$$

Choose a number

$$\Delta \ge \Delta_1 + 2b_* + 2. \tag{2.194}$$

Assume that

$$g \in \mathcal{U}, \; T_1 \in R^1, \; T_2 \ge T_1 + \Delta, \; \tau \in [T_1, T_2 - \Delta], \tag{2.195}$$

a trajectory-control pair

$$x : [T_1, T_2] \to R^n, \; u : [T_1, T_2] \to R^m$$

satisfies (2.186) and

$$\tau \in [T_1, T_2 - \Delta].$$

By (2.195), the choice of $\mathcal{U}_1$ and (2.189),

$$I^g(T_1, T_2, x, u) < \infty.$$

In order to complete the proof of the lemma it is sufficient to show that (2.187) holds.

Assume the contrary. Then

$$|x(t) - x_*(t)| \geq \epsilon_0 \text{ for all } t \in (\tau, \tau + \Delta). \tag{2.196}$$

It is not difficult to see that there exist real numbers τ_1, τ_2 such that

$$T_1 \leq \tau_1 < \tau_2 \leq T_2,$$

$$[\tau, \tau + \Delta] \subset [\tau_1, \tau_2], \tag{2.197}$$

$$|x(t) - x_*(t)| \geq \epsilon_0 \text{ for all } t \in [\tau_1, \tau_2] \tag{2.198}$$

and one of the following cases holds:

$$|x(\tau_i) - x_*(\tau_i)| = \epsilon_0, \ i = 1, 2; \tag{2.199}$$

$$|x(\tau_1) - x_*(\tau_1)| = \epsilon_0, \ \tau_2 = T_2; \tag{2.200}$$

$$|x(\tau_2) - x_*(\tau_2)| = \epsilon_0, \ \tau_1 = T_1; \tag{2.201}$$

$$\tau_i = T_i, \ i = 1, 2. \tag{2.202}$$

Define $\xi_1, \xi_2 \in R^n$ as follows:

$$\text{if (2.199) holds, then } \xi_i = x(\tau_i), \ 1, 2; \tag{2.203}$$

$$\text{if (2.200) holds, then } \xi_1 = x(\tau_1), \ \xi_2 = x_*(T_2); \tag{2.204}$$

$$\text{if (2.201) holds, then } \xi_2 = x(\tau_2), \ \xi_1 = x_*(T_1); \tag{2.205}$$

$$\text{if (2.202) holds, then } \xi_i = x_*(\tau_i), \ i = 1, 2. \tag{2.206}$$

We will define a trajectory-control pair

$$y : [\tau_1, \tau_2] \to R^n, \ v : [\tau_1, \tau_2] \to R^m.$$

Since property (iv) holds with $\epsilon = 1$ and $\delta = \epsilon_0$ it follows from (2.199)–(2.205) that there exist trajectory-control pairs

$$y_1 : [\tau_1, \tau_1 + b_*] \to R^n, \ v_1 : [\tau_1, \tau_1 + b_*] \to R^m$$

and

$$y_2 : [\tau_2 - b_*, \tau_2] \to R^n, \ v_2 : [\tau_2 - b_*, \tau_2] \to R^m$$

such that

$$y_1(\tau_1) = \xi_1,$$
$$y_1(\tau_1 + b_*) = x_*(\tau_1 + b_*), \tag{2.207}$$
$$|y_1(t) - x_*(t)| \le 1 \text{ for all } t \in [\tau_1, \tau_1 + b_*], \tag{2.208}$$
$$I^f(\tau_1, \tau_1 + b_*, y_1, v_1) \le I^f(\tau_1, \tau_1 + b_*, x_*, u_*) + 1, \tag{2.209}$$
$$y_2(\tau_2 - b_*) = x_*(\tau_2 - b_*),$$
$$y_2(\tau_2) = \xi_2, \tag{2.210}$$
$$|y_2(t) - x_*(t)| \le 1 \text{ for all } t \in [\tau_2 - b_*, \tau_2], \tag{2.211}$$
$$I^f(\tau_2 - b_*, \tau_2, y_2, v_2) \le I^f(\tau_2 - b_*, \tau_2, x_*, u_*) + 1. \tag{2.212}$$

By (2.208), (2.38), (2.39), (2.209), (2.211), and (2.212)

$$I^{f_r}(\tau_1, \tau_1 + b_*, y_1, v_1) \le I^f(\tau_1, \tau_1 + b_*, y_1, v_1) + b_*$$
$$\le I^f(\tau_1, \tau_1 + b_*, x_*, u_*) + 1 + b_* \tag{2.213}$$

and

$$I^{f_r}(\tau_2 - b_*, \tau_2, y_2, v_2) \le I^f(\tau_2 - b_*, \tau_2, y_2, v_2) + b_*$$
$$\le I^f(\tau_2 - b_*, \tau_2, x_*, u_*) + 1 + b_*. \tag{2.214}$$

By (2.197), (2.207), and (2.210) there exists a trajectory-control pair

$$y : [\tau_1, \tau_2] \to R^n, \ v : [\tau_1, \tau_2] \to R^m$$

such that

$$y(t) = y_1(t), \ v(t) = v_1(t) \text{ for all } t \in [\tau_1, \tau_1 + b_*],$$
$$y(t) = x_*(t), \ v(t) = u_*(t) \text{ for all } t \in (\tau_1 + b_*, \tau_2 - b_*),$$
$$y(t) = y_2(t), \ v(t) = v_2(t) \text{ for all } t \in [\tau_2 - b_*, \tau_2]. \tag{2.215}$$

By (2.215), (2.189), (2.195), and the choice of $\mathcal{U}_1$ [see (2.190), (2.191)],

$$I^g(\tau_1 + b_*, \tau_2 - b_*, y, v)$$

is finite.

By (2.215), (2.213), (2.214), and (2.189),

$$I^{fr}(\tau_1, \tau_1 + b_*, y, v) \leq I^f(\tau_1, \tau_1 + b_*, x_*, u_*) + 1 + b_* \leq \tilde{\Lambda} + 1 + b_*,$$
$$I^{fr}(\tau_2 - b_*, \tau_2, y, v) \leq I^f(\tau_2 - b_*, \tau_2, x_*, u_*) + 1 + b_* \leq \tilde{\Lambda} + 1 + b_*.$$

Together with (2.215) and the choice of $\mathcal{U}_1$ [see (2.190), (2.191)] this implies that

$$I^g(\tau_1, \tau_1 + b_*, y, v) \leq \tilde{\Lambda} + 2 + b_*,$$
$$I^g(\tau_2 - b_*, \tau_2, y, v) \leq \tilde{\Lambda} + 2 + b_*. \tag{2.216}$$

By the construction of (y, v) [see (2.215)], (2.207) and (2.210),

$$y(\tau_1) = \xi_1, \ y(\tau_2) = \xi_2. \tag{2.217}$$

By (2.186), (2.217) and the choice of ξ_1, ξ_2 [see (2.199)–(2.206)],

$$S \geq I^g(\tau_1, \tau_2, x, u) - I^g(\tau_1, \tau_2, y, v). \tag{2.218}$$

In view of (2.218), (2.215), (2.216), and (A),

$$S \geq I^g(\tau_1 + b_*, \tau_2 - b_*, x, u) - I^g(\tau_1 + b_*, \tau_2 - b_*, x_*, u_*)$$
$$+ I^g(\tau_1, \tau_1 + b_*, x, u) + I^g(\tau_2 - b_*, \tau_2, x, u)$$
$$- I^g(\tau_1, \tau_1 + b_*, y, v) - I^g(\tau_2 - b_*, \tau_2, y, v)$$
$$\geq I^g(\tau_1 + b_*, \tau_2 - b_*, x, u) - I^g(\tau_1 + b_*, \tau_2 - b_*, x_*, u_*)$$
$$- 2a_0 b_* - 2\tilde{\Lambda} - 4 - 2b_*$$

and

$$I^g(\tau_1 + b_*, \tau_2 - b_*, x, u) - I^g(\tau_1 + b_*, \tau_2 - b_*, x_*, u_*)$$
$$\leq S + 2b_*(1 + a_0) + 4 + 2\tilde{\Lambda}. \tag{2.219}$$

On the other hand it follows from (2.195), (2.194), (2.197), (2.198) and the choice of $\mathcal{U}$ [see (2.192), (2.193)] that

$$I^g(\tau_1 + b_*, \tau_2 - b_*, x, u) > I^g(\tau_1 + b_*, \tau_2 - b_*, x_*, u_*)$$
$$+ S + 2b_*(1 + a_0) + 8 + 2\tilde{\Lambda}.$$

This contradicts (2.219). The contradiction we have reached completes the proof of Lemma 2.23.   □

We suppose that the sum over empty set is zero.

Lemma 2.24. *Let $S, \epsilon > 0$. Then there exist real numbers $l > 0$, $\Delta > 0$, a natural number Q, and a neighborhood $\mathcal{U}$ of f_r in $\mathfrak{M}$ such that for each $g \in \mathcal{U}$, each $T_1 \in R^1$, each $T_2 \geq T_1 + \Delta$, and each trajectory-control pair*

$$x : [T_1, T_2] \to R^n, \ u : [T_1, T_2] \to R^m$$

which satisfies

$$I^g(T_1, T_2, x, u) \leq \inf\{\sigma^g(T_1, T_2, y) : \ (T_1, y) \in A\} + S \quad (2.220)$$

there exist finite sequences

$$\{a_i\}_{i=1}^q, \ \{b_i\}_{i=1}^q \subset [T_1, T_2],$$

where $q \leq Q$ is a natural number, such that

$$a_i \leq b_i \leq a_i + l \text{ for all integers } i = 1, \ldots, q$$

and

$$\{t \in [T_1, T_2] : \ |x(t) - x_*(t)| > \epsilon\}$$
$$\subset \cup_{i=1}^q [a_i, b_i].$$

Proof. We may assume without loss of generality that

$$\epsilon < \min\{1, d_0\}. \quad (2.221)$$

By Lemma 2.19, there exist

$$\delta_0 \in (0, \epsilon) \text{ and } \Delta_1 \geq 1,$$

and a neighborhood $\mathcal{U}_1$ of f_r in $\mathfrak{M}$ such that for each $g \in \mathcal{U}_1$, each $T_1 \in R^1$, each $T_2 \geq T_1 + \Delta_1$, and each trajectory-control pair

$$x : [T_1, T_2] \to R^n, \ u : [T_1, T_2] \to R^m$$

which satisfies

$$|x(T_i) - x_*(T_i)| \leq \delta_0, \ i = 1, 2, \quad (2.222)$$

$$I^g(T_1, T_2, x, u) \leq U^g(T_1, T_2, x(T_1), x(T_2)) + \delta_0 \quad (2.223)$$

the following inequality holds:

$$|x(t) - x_*(t)| \leq \epsilon \text{ for all } t \in [T_1, T_2]. \tag{2.224}$$

By Lemma 2.23, there exist $\Delta_2 > 0$ and a neighborhood $\mathcal{U}_2$ of f_r in $\mathfrak{M}$ such that for each $g \in \mathcal{U}_2$, each pair of real numbers

$$T_1 \in R^1, \; T_2 \geq T_1 + \Delta_2,$$

each trajectory-control pair

$$x : [T_1, T_2] \to R^n, \; u : [T_1, T_2] \to R^m$$

which satisfies (2.220) and each

$$\tau \in [T_1, T_2 - \Delta_2],$$

we have

$$\min\{|x(t) - x_*(t)| : \; t \in (\tau, \tau + \Delta_2)\} < \delta_0. \tag{2.225}$$

Set

$$\mathcal{U} = \mathcal{U}_1 \cap \mathcal{U}_2, \tag{2.226}$$

$$l = 2\Delta_2 + 2\Delta_1 \tag{2.227}$$

and choose

$$\Delta \geq 8\Delta_1 + 8\Delta_2 \tag{2.228}$$

and a natural number

$$Q > 6 + 2\delta_0^{-1} S. \tag{2.229}$$

Assume that

$$g \in \mathcal{U}, \; T_1 \in R^1, \; T_2 \geq T_1 + \Delta \tag{2.230}$$

and that a trajectory-control pair

$$x : [T_1, T_2] \to R^n, \; u : [T_1, T_2] \to R^m$$

satisfies (2.220). By (2.230), (2.226), (2.228), (2.220), and the choice of $\mathcal{U}_2$ [see (2.225)], there exists a strictly increasing sequence of real numbers $\{t_i\}_{i=1}^p$, where p is a natural number, such that

$$\{t_i\}_{i=1}^p \subset [T_1, T_2],$$
$$|x(t_i) - x_*(t_i)| < \delta_0, \quad i = 1, \ldots, p, \tag{2.231}$$
$$t_1 \in [T_1, T_1 + \Delta_2],$$
$$\Delta_2 \leq t_{i+1} - t_i \leq 2\Delta_2$$

for each integer i satisfying $1 \leq i < p$,

$$t_p \geq T_2 - 2\Delta_2. \tag{2.232}$$

Set

$$S_1 = t_1. \tag{2.233}$$

By induction we construct finite strictly increasing sequences of numbers

$$\{r(i)\}_{i=1}^k \subset \{1, \ldots, p\},$$
$$\{S_i\}_{i=1}^k \subset \{t_i : i = 1, \ldots, p\}$$

such that

$$S_i = t_{r(i)}, \quad i = 1, \ldots, k, \tag{2.234}$$
$$r(k) = p \text{ and } S_k = t_p \tag{2.235}$$

such that the following two properties hold:

(C1) For each integer i satisfying $1 \leq i < k - 1$,

$$I^g(S_i, S_{i+1}, x, u) > U^g(S_i, S_{i+1}, x(S_i), x(S_{i+1})) + \delta_0; \tag{2.236}$$

(C2) If an integer i satisfies $1 \leq i \leq k - 1$, (2.236) holds and

$$r(i + 1) > r(i) + 1,$$

then

$$I^g(S_i, t_{r(i+1)-1}, x, u) \leq U^g(S_i, t_{r(i+1)-1}, x(S_i), x(t_{r(i+1)-1})) + \delta_0. \tag{2.237}$$

Assume that an integer $j \geq 1$ and we have already defined strictly increasing sequences of numbers

$$\{r(i)\}_{i=1}^j \subset \{1, \ldots, p\},$$
$$\{S_i\}_{i=1}^j \subset \{t_i : i = 1, \ldots, p\}$$

such that

$$S_i = t_{r(i)}, \quad i = 1, \ldots, j,$$
$$S_j < t_p,$$

for each integer i satisfying $1 \leq i < j$ relation (2.236) holds, and if an integer i satisfies $1 \leq i < j$ and

$$r(i + 1) > r(i) + 1,$$

then (2.237) is true. (Clearly, for $j = 1$ our assumption holds.)
   Let us define

$$r(j + 1) \in \{1, \ldots, p\} \text{ and } S_{j+1} = t_{r(j+1)}.$$

If

$$I^g(S_j, t_p, x, u) \leq U^g(S_j, t_p, x(S_j), x(t_p)) + \delta_0,$$

then we set

$$r(j + 1) = p, \quad S_{j+1} = t_p,$$

complete our construction and it is easy to see that properties (C1) and (C2) hold.
   Assume that

$$I^g(S_j, t_p, x, u) > U^g(S_j, t_p, x(S_j), x(t_p)) + \delta_0. \tag{2.238}$$

Set

$$r(j + 1) = \min\{i \in \{r(j) + 1, \ldots, p\} :$$
$$I^g(S_j, t_i, x, u) > U^g(S_j, t_i, x(S_j), x(t_i)) + \delta_0\}. \tag{2.239}$$

If $r(j + 1) = p$, then we set $k = j + 1$, complete our construction and it is easy to see that properties (C1) and (C2) hold.
   If $r(j + 1) < p$, then we set $S_{j+1} = t_{r(j+1)}$ and it is easy to see that the assumption made for j also holds for $j+1$. Clearly, our construction of the sequences is completed after a finite number of steps. Let $r(k)$ and S_k be their last elements respectively. It follows from our construction that $r(k) = p$, $S_k = t_p$ and that (C1 and (C2) hold.

By (2.220) and (C1),

$$S \geq I^g(T_1, T_2, x, u) - \inf\{\sigma^g(T_1, T_2, y): (T_1, y) \in A\}$$

$$\geq \sum \{I^g(S_i, S_{i+1}, x, u) - U^f(S_i, S_{i+1}, x(S_i), x(S_{i+1})):$$

$$i \text{ is an integer such that } 1 \leq i < k - 1\} \geq (k - 2)\delta_0$$

and

$$k \leq 2 + \delta_0^{-1}S. \tag{2.240}$$

Set

$$A = \{i \in \{1, \ldots, k\} : i < k \text{ and}$$
$$S_{i+1} - S_i > 2\Delta_2 + 2\Delta_1\}. \tag{2.241}$$

Let

$$i \in A. \tag{2.242}$$

By (2.234), (2.242), and (2.241),

$$t_{r(i+1)} - t_{r(i)} = S_{i+1} - S_i > 2\Delta_2 + 2\Delta_1. \tag{2.243}$$

By (2.243) and (2.232),

$$r(i + 1) > r(i) + 1. \tag{2.244}$$

By (2.242), (2.241), (2.244), (C1) and (C2),

$$I^g(t_{r(i)}, t_{r(i+1)-1}, x, u) \leq U^g(t_{r(i)}, t_{r(i+1)-1}, x(t_{r(i)}), x(t_{r(i+1)-1})) + \delta_0. \tag{2.245}$$

By (2.244), (2.232), and (2.243),

$$t_{r(i+1)-1} - t_{r(i)} \geq t_{r(i+1)} - 2\Delta_2 - t_{r(i)} \geq 2\Delta_1. \tag{2.246}$$

By (2.246), (2.245), (2.231), and (2.230), (2.226) and the choice of $\mathcal{U}_1$, δ_0, Δ_1 [see (2.222)–(2.224)],

$$|x(t) - x_*(t)| \leq \epsilon \text{ for all } t \in [t_{r(i)}, t_{r(i+1)-1}]. \tag{2.247}$$

In view of (2.247), (2.234), and (2.232),

$$|x(t) - x_*(t)| \leq \epsilon \text{ for all } t \in [S_i, S_{i+1} - 2\Delta_2] \tag{2.248}$$

for all $i \in A$. By (2.248),

$$\{t \in [T_1, T_2] : |x(t) - x_*(t)| > \epsilon\}$$

$$\subset (\cup\{[S_i, S_{i+1}] : i \in \{1, \ldots, k-1\} \setminus A\}) \cup \{\cup[S_{i+1} - 2\Delta_2, S_{i+1}] : i \in A\}$$

$$\cup [T_1, t_1] \cup [t_p, T_2]. \tag{2.249}$$

The right-hand side of (2.249) is a finite union of closed intervals. By (2.240) and (2.229) their number does not exceed

$$2k + 2 \leq 6 + 2\delta_0^{-1} S \leq Q$$

and in view of (2.241), (2.232), and (2.227) their maximal length does not exceed

$$2\Delta_2 + 2\Delta_1 = l.$$

Lemma 2.24 is proved.   □

2.5   Proofs of Theorems 2.1–2.3 and 2.5

Proof of Theorem 2.1. The validity of assertion 1 follows from Lemma 2.10 and assumptions (A) and (B).

We will prove assertion 2. Let $f \in \mathfrak{M}_{reg}$, $s \in R^1$ and

$$x : [s, \infty) \to R^n, \ u : [s, \infty) \to R^m$$

be a trajectory-control pair. Assume that there exists a sequence of numbers $\{t_k\}_{k=1}^\infty$ such that

$$t_k \to \infty \text{ as } k \to \infty, \tag{2.250}$$

$$I^f(s, t_k, x, u) - I^f(s, t_k, x_f, u_f) \to \infty \text{ as } k \to \infty. \tag{2.251}$$

We will show that relation (a) holds. It follows from assumption (A) and (B) that there exists a number $S_0 > 0$ such that

$$S_0 > 2|x(s)| + 8,$$

$$\psi(S_0 - 4) - a_0 - 8 \geq 8 \sup\{|I^f(j, j+1, x_f, u_f)| : j = 0, \pm 1, \pm 2, \ldots\}. \tag{2.252}$$

Let a number $S > 0$ be as guaranteed in assertion 1. By assumption (A) and (2.252) we may assume without loss of generality that

$$\liminf_{t \to \infty} |x(t)| \leq S_0 - 1. \tag{2.253}$$

For each integer $k \geq 1$ we set

$$\tau_k = \inf\{t \in [t_k, \infty): |x(t)| \leq S_0\}. \qquad (2.254)$$

Let $k \geq 1$ be an integer and $t \geq \tau_k$. It follows from (2.254), (2.251), (2.252), the choice of S, assertion 1, and assumption (A) that

$$I^f(s, t, x, u) - I^f(s, t, x_f, u_f)$$
$$\geq I^f(s, t_k, x, u) - I^f(s, t_k, x_f, u_f)$$
$$+ I^f(t_k, \tau_k, x, u) - I^f(t_k, \tau_k, x_f, u_f) - S$$
$$\geq I^f(s, t_k, x, u) - I^f(s, t_k, x_f, u_f) - S - 4a_0$$
$$-2 \sup\{|I^f(j, j+1, x_f, u_f)|: j = 0, \pm1, \pm2, \ldots\} \to \infty \text{ as } k \to \infty.$$

Therefore relation (a) holds. Together with assertion 1 and Proposition 2.7 this implies the validity of assertion 2. □

Construction of the set $\mathcal{F}$:

Let $\mathfrak{A}$ be a subset of $\mathfrak{M}_{reg}$ such that $f_r \in \mathfrak{A}$ for each $f \in \mathfrak{A}$ each $r \in (0, 1)$. Denote by $\bar{\mathfrak{A}}$ the closure of $\mathfrak{A}$ in the space $\mathfrak{M}$.

It is easy to see that for each $f \in \mathfrak{A}$ and each $r \in (0, 1)$ Lemmas 2.12–2.24 hold with

$$x_* = x_f, \ u_* = u_f, \ b_* = b_f, \ d_0 = 1.$$

Relation (2.31) implies that

$$\sup\{|x_f(t)|: t \in R^1\} < \infty \text{ for each } f \in \mathfrak{A}. \qquad (2.255)$$

Set

$$E = \{f_r: f \in \mathfrak{A}, \ r \in (0, 1)\}.$$

It follows from Lemma 2.11 that E is an everywhere dense subset of $\bar{\mathfrak{A}}$.

For each $f \in \mathfrak{A}$, each $r \in (0, 1)$, and each integer $k \geq 1$ there exist an open neighborhood $V(f, r, k)$ of f_r and numbers

$$\delta(f, r, k) \in (0, (2k)^{-1}),$$
$$\gamma(f, r, k) \in (0, \delta(f, r, k)),$$
$$\Delta(f, r, k), \ Q(f, r, k) \geq 1,$$
$$l(f, r, k) > 0, \ L(f, r, k) > 0$$

and a natural number $p(f, r, k)$ such that:

(a) Lemma 2.20 holds for f and r with

$$S = k, \ \epsilon = (2k)^{-1}, \ \mathcal{U} = V(f, r, k),$$
$$\Delta = \Delta(f, r, k), \ \delta = 2\delta(f, r, k),$$
$$x_* = x_f, \ u_* = u_f;$$

(b) Lemma 2.21 holds for f and r with $\mathcal{U} = V(f, r, k)$;
(c) Lemma 2.22 holds for f and r with

$$S = k, \ \mathcal{U} = V(f, r, k),$$
$$\delta = \delta(f, r, k), \ Q = Q(f, r, k);$$

(d) Lemma 2.20 holds for f and r with

$$S = k, \ \epsilon = \delta(f, r, k), \ \mathcal{U} = V(f, r, k),$$
$$\Delta = \Delta(f, r, k), \ \delta = \gamma(f, r, k),$$
$$x_* = x_f, \ u_* = u_f;$$

(e) Lemma 2.24 holds for f and r with

$$S = 4k, \ \epsilon = (4k)^{-1}, \ \mathcal{U} = V(f, r, k),$$
$$l = l(f, r, k), \ \Delta = L(f, r, k), \ Q = p(f, r, k),$$
$$x_* = x_f, \ u_* = u_f.$$

We define

$$\mathcal{F} = \bar{\mathfrak{A}} \cap [\cap_{k=1}^{\infty} \cup \{V(f, r, k) : \ f \in \mathfrak{A}, \ r \in (0, 1)\}]. \qquad (2.256)$$

Clearly, $\mathcal{F}$ is a countable intersection of open everywhere dense sets in $\bar{\mathfrak{A}}$.

Theorem 2.2 now follows from the definition of $\mathcal{F}$, properties (b), (c) and Lemmas 2.21 and 2.22.

Proof of Theorem 2.3. Let $f \in \mathcal{F}$. For each integer $k \geq 1$ there exist

$$f_k \in \mathfrak{A} \text{ and } r_k \in (0, 1)$$

such that

$$f \in V(f_k, r_k, k). \qquad (2.257)$$

Let p, q be natural numbers. We show that

$$|x_{f_p}(t) - x_{f_q}(t)| \le \delta(f_p, r_p, p) + \delta(f_q, r_q, q) \text{ for all } t \in R^1. \qquad (2.258)$$

Let N be a natural number. Set

$$T_2 = 4N + 4 + 4\Delta(f_p, r_p, p) + 4\Delta(f_q, r_q, q),$$
$$T_1 = -T_2. \qquad (2.259)$$

It follows from Theorem 2.2 that

$$\inf\{U^f(T_1, T_2, y_1, y_2) : (T_i, y_i) \in A, \ i = 1, 2\} < \infty. \qquad (2.260)$$

Therefore there exists a trajectory-control pair

$$x : [T_1, T_2] \to R^n, \ u : [T_1, T_2] \to R^m$$

such that

$$I^f(T_1, T_2, x, u) \le \inf\{U^f(T_1, T_2, y_1, y_2) : (T_i, y_i) \in A, \ i = 1, 2\}$$
$$+ 8^{-1} \min\{\gamma(f_p, r_p, p), \ \gamma(f_q, r_q, q)\}. \qquad (2.261)$$

By property (d), for each integer $k \ge 1$, Lemma 2.20 holds with

$$f = f_k, \ r = r_k, \ S = k,$$
$$\epsilon = \delta(f_k, r_k, k), \ \mathcal{U} = V(f_k, r_k, k),$$
$$\Delta = \Delta(f_k, r_k, k), \ \delta = \gamma(f_k, r_k, k),$$
$$x_* = x_{f_k}, \ u_* = u_{f_k}.$$

Together with (2.260), (2.261), (2.259), and (2.257) this implies that

$$|x(t) - x_{f_p}(t)| \le \delta(f_p, r_p, p), \ t \in [-N, N],$$
$$|x(t) - x_{f_q}(t)| \le \delta(f_q, r_q, q), \ t \in [-N, N]$$

and

$$|x_{f_q}(t) - x_{f_p}(t)| \le \delta(f_p, r_p, p) + \delta(f_q, r_q, q), \ t \in [-N, N].$$

Since this relation holds for any integer $N \ge 1$ we conclude that (2.258) holds. It follows from (2.258) and (2.255) that there exists a bounded continuous function $X_f : R^1 \to R^n$ which satisfies

$$|X_f(t) - x_{f_p}(t)| \le \delta(f_p, r_p, p) \text{ for all } t \in R^1 \text{ and all integers } p \ge 1. \qquad (2.262)$$

Let $S, \epsilon > 0$. Fix an integer

$$p > 4 + 4S + 8\epsilon^{-1} \tag{2.263}$$

and set

$$\mathcal{U} = V(f_p, r_p, p),$$
$$\Delta = \Delta(f_p, r_p, p), \ \delta = \delta(f_p, r_p, p). \tag{2.264}$$

Assume that

$$g \in \mathcal{U}, \ T_1 \in R^1, \ T_2 \geq T_1 + 2\Delta \tag{2.265}$$

and a trajectory-control pair

$$x : [T_1, T_2] \to R^n, \ u : [T_1, T_2] \to R^m$$

such that

$$I^g(T_1, T_2, x, u) \leq \inf\{U^g(T_1, T_2, y_1, y_2) : \ (T_i, y_i) \in A, \ i = 1, 2\} + S,$$
$$I^g(T_1, T_2, x, u) \leq U^g(T_1, T_2, x(T_1), x(T_2)) + \delta. \tag{2.266}$$

It follows from (2.264), (2.265), (2.266), (2.263), and Lemma 4.20 which holds with

$$f = f_p, \ r = r_p, \ S = p,$$
$$\epsilon = (2p)^{-1}, \ \mathcal{U} = V(f_p, r_p, p),$$
$$\Delta = \Delta(f_p, r_p, p), \ \delta = 2\delta(f_p, r_p, p),$$
$$x_* = x_{f_p}, \ u_* = u_{f_p}$$

that

$$|x(t) - x_{f_p}(t)| \leq p^{-1}, \ t \in [T_1 + \Delta, T_2 - \Delta].$$

Moreover, if $|x(T_1) - x_{f_p}(T_1)| \leq \delta$, then

$$|x(t) - x_{f_p}(t)| \leq p^{-1}, \ t \in [T_1, T_2 - \Delta]$$

and if $|x(T_2) - x_{f_p}(T_2)| \leq \delta$, then

$$|x(t) - x_{f_p}(t)| \leq p^{-1}, \ t \in [T_1 + \Delta, T_2].$$

Together with (2.262) and (2.263) this implies the validity of Theorem 2.3. $\square$

Proof of Theorem 2.5. Choose a natural number k such that

$$k > 4 + 4M + 8\epsilon^{-1}. \tag{2.267}$$

By (2.256) there exist

$$f_k \in \mathfrak{A}, \ r_k \in (0,1)$$

such that

$$f \in V(f_k, r_k, k). \tag{2.268}$$

Set

$$l = l(f_k, r_k, k), \ L = L(f_k, r_k, k), \ p = p(f_k, r_k, k),$$
$$\mathcal{U} = V(f_k, r_k, k). \tag{2.269}$$

Assume that

$$g \in \mathcal{U}, \ T_1 \in R^1, \ T_2 \geq T_1 + L \tag{2.270}$$

and a trajectory-control pair

$$x : [T_1, T_2] \to R^n, \ u : [T_1, T_2] \to R^m$$

satisfies

$$I^g(T_1, T_2, x, u) \leq \inf\{U^g(T_1, T_2, y_1, y_2) : (T_i, y_i) \in A, \ i = 1, 2\} + M. \tag{2.271}$$

By (2.270), (2.271), (2.269), (2.267), condition (e), and Lemma 2.24 which holds with

$$\epsilon = (4k)^{-1}, \ S = 4k,$$
$$f = f_k, \ r = r_k,$$
$$\mathcal{U} = V(f_k, r_k, k),$$
$$x_* = x_{f_k}, \ u_* = u_{f_k},$$
$$l = l(f_k, r_k, k), \ \Delta = L(f_k, r_k, k), \ Q = p(f_k, r_k, k)$$

that there exist finite sequences of numbers

$$\{a_i\}_{i=1}^q, \ \{b_i\}_{i=1}^q \subset [T_1, T_2],$$

where q is a natural number, such that

$$q \leq p(f_k, r_k, k) = p, \tag{2.272}$$

$$a_i \leq b_i \leq a_i + l \text{ for all integers } i = 1, \ldots, q \tag{2.273}$$

and

$$\{t \in [T_1, T_2] : |x(t) - x_{f_k}(t)| > (4k)^{-1}\}$$
$$\subset \cup_{i=1}^{q} [a_i, b_i]. \tag{2.274}$$

By Theorem 2.3 there exist

$$\delta_0 \in (0, 1), \ \tau_0 > 0$$

such that for each $\tau \geq \tau_0$ and each trajectory-control pair

$$y : [-\tau, \tau] \to R^n, \ v : [-\tau, \tau] \to R^m$$

satisfying

$$I^f(-\tau, \tau, y, v) \leq \inf\{U^f(-\tau, \tau, \xi_1, \xi_2) : (-\tau, \xi_1), (\tau, \xi_2) \in A\} + \delta_0 \tag{2.275}$$

we have

$$|y(t) - X_f(t)| \leq (8k)^{-1}, \ t \in [-\tau + \tau_0, \tau - \tau_0]. \tag{2.276}$$

Let

$$T > \tau_0 + \Delta(f_k, r_k, k) \tag{2.277}$$

and a trajectory-control pair

$$y : [-T, T] \to R^n, \ v : [-T, T] \to R^m$$

satisfy

$$I^f(-T, T, y, v) \leq \inf\{U^f(-T, T, \xi_1, \xi_2) : (-T, \xi_1), (T, \xi_2) \in A\}$$
$$+ \min\{\delta_0, \ \delta(f_k, r_k, k)\}. \tag{2.278}$$

By (2.277), (2.278) and the choice of δ_0, τ_0 [see (2.275), (2.276)],

$$|y(t) - X_f(t)| \leq (8k)^{-1}, \ t \in [-T + \tau_0, T - \tau_0]. \tag{2.279}$$

By (2.277), (2.278), (2.268), condition (a), Lemma 2.20 which holds for f_k and r_k with

$$S = k, \; \epsilon = (2k)^{-1}, \; \mathcal{U} = V(f_k, r_k, k),$$
$$\Delta = \Delta(f_k, r_k, k), \; \delta = 2\delta(f_k, r_k, k),$$
$$x_* = x_{f_k}, \; u_* = u_{f_k}$$

we have

$$|y(t) - x_{f_k}(t)| \le (2k)^{-1} \text{ for all } t \in [-T + \Delta(f_k, r_k, k), T - \Delta(f_k, r_k, k)].$$

Together with (2.279) this implies that

$$|X_f(t) - x_{f_k}(t)| \le (8k)^{-1} + (2k)^{-1}$$
$$\text{for all } t \in [-T + \tau_0 + \Delta(f_k, r_k, k), T - \Delta(f_k, r_k, k) - \tau_0].$$

Since T is any natural number satisfying (2.277) we conclude that

$$|X_f(t) - x_{f_k}(t)| \le (8k)^{-1} + (2k)^{-1} \text{ for all } t \in R^1.$$

Together with (2.274) and (2.267) this implies that for all

$$t \in [T_1, T_2] \setminus \cup_{i=1}^q [a_i, b_i],$$
$$|x(t) - X_f(t)| \le |x(t) - x_{f_k}(t)| + |x_{f_k}(t) - X_f(t)|$$
$$\le (4k)^{-1} + (8k)^{-1} + (2k)^{-1} < k^{-1} < \epsilon.$$

Theorem 2.5 is proved. $\square$

Chapter 3
Infinite Horizon Problems

In this chapter we continue to use the notation and the definitions of Chap. 2.

3.1 Existence of Optimal Solutions

Let $f \in \mathfrak{M}$ and τ be a real number. A trajectory-control pair

$$x : [\tau, \infty) \to R^n, \ u : [\tau, \infty) \to R^m$$

is (f)-*overtaking optimal* if for any trajectory-control pair

$$y : [\tau, \infty) \to R^n, \ v : [\tau, \infty) \to R^m$$

satisfying

$$y(\tau) = x(\tau)$$

the following inequality holds:

$$\limsup_{T \to \infty}[I^f(\tau, T, x, u) - I^f(\tau, T, y, v)] \leq 0.$$

This notion, known as the overtaking optimality criterion, was introduced in the economics literature [17,43] and has been used in optimal control theory [12,24,49, 50].

A trajectory-control pair

$$x : I \to R^n, \ u : I \to R^m,$$

where I is either R^1 or $[T, \infty)$ (with $T \in R^1$) is (f)-good [12, 17, 50] if there exists a number $S > 0$ such that for each $T_1, T_2 \in I$ satisfying $T_2 > T_1$,

$$I^f(T_1, T_2, x, u) \leq \inf\{U^f(T_1, T_2, y_1, y_2) : (T_i, y_i) \in A, \ i = 1, 2\} + S.$$

Remark 3.1. It follows from Theorem 2.1 and assumptions (A) and (B) that

$$x_f : R^1 \to R^n, \ u_f : R^1 \to R^m$$

is an (f)-good trajectory-control pair for each $f \in \mathfrak{M}_{reg}$.

We say that a function $f \in \mathfrak{M}$ has the *turnpike property* if there exists a bounded continuous function $X_f : R^1 \to R^n$ such that:
for each $S, \epsilon > 0$ there exist numbers $\Delta, \delta > 0$ such that for each pair of numbers

$$T_1 \in R^1, \ T_2 \geq T_1 + 2\Delta$$

and each trajectory-control pair

$$x : [T_1, T_2] \to R^n, \ u : [T_1, T_2] \to R^m$$

which satisfies

$$I^f(T_1, T_2, x, u) \leq \inf\{U^f(T_1, T_2, y_1, y_2) : (T_i, y_i) \in A, \ i = 1, 2\} + S$$

and

$$I^f(T_1, T_2, x, u) \leq U^f(T_1, T_2, x(T_1), x(T_2)) + \delta$$

the following inequality holds:

$$|x(t) - X_f(t)| \leq \epsilon \text{ for all } t \in [T_1 + \Delta, T_2 - \Delta].$$

We can easily deduce the following result.

Proposition 3.2. *Assume that $f \in \mathfrak{M}$ has the turnpike property, $\tau \in R^1$ and that*

$$x : [\tau, \infty) \to R^n, \ u : [\tau, \infty) \to R^m$$

is an (f)-good trajectory-control pair. Then

$$x(t) - X_f(t) \to 0 \text{ as } t \to \infty.$$

The following optimality criterion for infinite horizon problems was introduced by Aubry and Le Daeron [7] in their study of the discrete Frenkel–Kontorova model related to dislocations in one-dimensional crystals.

Let $f \in \mathfrak{M}$. A trajectory-control pair

$$x : I \to R^n, \ u : I \to R^m,$$

where I is either R^1 or $[T_1, \infty)$ or $[T_1, T_2]$ (with $-\infty < T_2 < T_2 < \infty$) is (f)-minimal if

$$I^f(T_1, T_2, x, u) = U^f(T_1, T_2, x(T_1), x(T_2))$$

for each pair of numbers $T_1, T_2 \in I$ satisfying $T_1 < T_2$.

We say that a function $f \in \mathfrak{M}$ has an *(LSC) property* if for each $T_1 \in R^1$, each $T_2 > T_1$, and each sequence of trajectory-control pairs

$$x_j : [T_1, T_2] \to R^n, \ u_j : [T_1, T_2] \to R^m, \ j = 1, 2, \dots$$

which satisfies

$$\sup\{I^f(T_1, T_2, x_j, u_j) : \ j = 1, 2, \dots\} < \infty$$

there exists a subsequence $\{(x_{j_k}, u_{j_k})\}_{k=1}^\infty$ and a trajectory-control pair

$$x : [T_1, T_2] \to R^n, \ u : [T_1, T_2] \to R^m$$

such that

$$x_{j_k}(t) \to x(t) \text{ as } k \to \infty \text{ for any } t \in [T_1, T_2]$$

and

$$I^f(T_1, T_2, x, u) \leq \liminf_{j \to \infty} I^f(T_1, T_2, x_j, u_j).$$

In Sect. 3.3 we will prove the following result.

Proposition 3.3. *Assume that $f \in \mathfrak{M}_{reg}$ has the (LSC) property and the turnpike property, $\tau \in R^1$ and that*

$$x : [\tau, \infty) \to R^n, \ u : [\tau, \infty) \to R^m$$

is an (f)-good trajectory-control pair. Then there exists an (f)-good and (f)-minimal trajectory-control pair

$$x_* : [\tau, \infty) \to R^n, \ u_* : [\tau, \infty) \to R^m$$

such that $x(\tau) = x_*(\tau)$.

The next result will be proved in Sect. 3.4.

Proposition 3.4. *Assume that $f \in \mathfrak{M}_{reg}$ has the (LSC) property and the turnpike property, $\tau \in R^1$ and that*

$$x : [\tau, \infty) \to R^n, \ u : [\tau, \infty) \to R^m$$

is an (f)-good and (f)-minimal trajectory-control pair. Then (x, u) is an (f)-overtaking optimal trajectory-control pair.

Propositions 3.3 and 3.4 imply the following result.

Theorem 3.5. *Assume that $f \in \mathfrak{M}_{reg}$ has the (LSC) property and the turnpike property, $\tau \in R^1$ and that*

$$x : [\tau, \infty) \to R^n, \ u : [\tau, \infty) \to R^m$$

is an (f)-good trajectory-control pair. Then there exists an (f)-overtaking optimal trajectory-control pair

$$x_* : [\tau, \infty) \to R^n, \ u_* : [\tau, \infty) \to R^m$$

such that $x(\tau) = x_*(\tau)$.

Let $f \in \mathfrak{M}$ and $\tau \in R^1$. A trajectory-control pair

$$x : [\tau, \infty) \to R^n, \ u : [\tau, \infty) \to R^m$$

is called (f)-*agreeable* [19–21, 52] if for each $T_0 > \tau$ and each $\epsilon > 0$ there exists $T_\epsilon > T_0$ such that for each $T > T_\epsilon$ there exists a trajectory-control pair

$$y : [\tau, T] \to R^n, \ v : [\tau, T] \to R^m$$

such that

$$y(t) = x(t), \ v(t) = u(t) \text{ for all } t \in [\tau, T_0]$$

and

$$I^f(\tau, T, y, v) \leq \sigma^f(\tau, T, x(\tau)) + \epsilon.$$

The following theorem is proved in Sect. 3.5.

Theorem 3.6. *Assume that $f \in \mathfrak{M}_{reg}$ has the turnpike property, $\tau \in R^1$ and that*

$$x : [\tau, \infty) \to R^n, \ u : [\tau, \infty) \to R^m$$

is an (f)-good trajectory-control pair. Then the trajectory-control pair (x, u) is (f)-minimal if and only if (x, u) is (f)-agreeable.

Let $f \in \mathfrak{M}$, $M > 0$ and $\tau \in R^1$. A point $\xi \in R^n$ is called (f, τ, M)-*good* if there exists a trajectory-control pair

$$x : [\tau, \infty) \to R^n, \ u : [\tau, \infty) \to R^m$$

such that

$$x(\tau) = \xi$$

and for each $T_1 \geq \tau$ and each $T_2 > T_1$,

$$I^f(T_1, T_2, x, u) \leq \inf\{U^f(T_1, T_2, \eta_1, \eta_2) : (T_i, \eta_i) \in A, \ i = 1, 2\} + M.$$

The following two turnpike results are proved in Sects. 3.6 and 3.7 respectively.

Theorem 3.7. *Assume that $f \in \mathfrak{M}_{reg}$ has the turnpike property, $M > 0$ and $\epsilon > 0$. Then there exists $L_0 > 0$ such that for each $\tau \in R^1$ and each (f)-good and (f)-minimal trajectory-control pair*

$$x : [\tau, \infty) \to R^n, \ u : [\tau, \infty) \to R^m$$

for which $x(\tau)$ is an (f, τ, M)-good point the following inequality holds:

$$|x(t) - x_f(t)| \leq \epsilon \text{ for all } t \geq \tau + L_0.$$

Theorem 3.8. *Assume that $f \in \mathfrak{M}_{reg}$ has the turnpike property, $M > 0$ and $\epsilon > 0$. Then there exists $l > 0$, $L > 0$ and a natural number Q such that for each $T_1 \in R^1$, each $T_2 \geq T_1 + L$ and each trajectory-control pair*

$$x : [T_1, T_2] \to R^n, \ u : [T_1, T_2] \to R^m$$

which satisfies

$$I^f(T_1, T_2, x, u) \leq \inf\{U^f(T_1, T_2, \xi_1, \xi_2) : (T_i, \xi_i) \in A, \ i = 1, 2\} + M$$

there exist finite sequences

$$\{a_i\}_{i=1}^q, \ \{b_i\}_{i=1}^q \subset [T_1, T_2],$$

where $q \leq Q$ is a natural number, such that

$$a_i \leq b_i \leq a_i + l \text{ for all integers } i = 1, \ldots, q$$

and

$$|x(t) - x_f(t)| \leq \epsilon \text{ for all } t \in [T_1, T_2] \setminus \cup_{i=1}^q [a_i, b_i].$$

3.2   Auxiliary Results

Let $f \in \mathfrak{M}_{reg}$ and x_f, u_f, b_f be as guaranteed by assumption (B).

Lemma 3.9. *Let $\epsilon_0 > 0$. Then there exists a number $\delta > 0$ such that for each* $(T_i, \xi_i) \in A$, $i = 1, 2$ *which satisfy*

$$T_2 - T_1 \geq 2b_f,$$
$$|\xi_i - x_f(T_i)| \leq \delta, \ i = 1, 2$$

the following relation holds:

$$U^f(T_1, T_2, \xi_1, \xi_2) \leq I^f(T_1, T_2, x_f, u_f) + \epsilon_0 < \infty.$$

Proof. Let

$$\epsilon = \epsilon_0/2 \tag{3.1}$$

and let $\delta > 0$ be as guaranteed by B(iv). Let

$$T_1 \in R^1, \ T_2 \geq T_1 + 2b_f \tag{3.2}$$

and $\xi_1, \xi_2 \in R^n$ be such that

$$(\xi_i, T_i) \in A, \ i = 1, 2 \tag{3.3}$$

and

$$|\xi_i - x_f(T_i)| \leq \delta, \ i = 1, 2. \tag{3.4}$$

By (3.2)–(3.4) and B(iv) which holds with ϵ and δ there exist trajectory-control pairs

$$x_1 : [T_1, T_1 + b_f] \to R^n, \ u_1 : [T_1, T_1 + b_f] \to R^m,$$
$$x_2 : [T_2 - b_f, T_2] \to R^n, \ u_2 : [T_2 - b_f, T_2] \to R^m$$

such that

$$x_1(T_1) = \xi_1,$$
$$x_1(T_1 + b_f) = x_f(T_1 + b_f),$$
$$x_2(T_2) = \xi_2,$$
$$x_2(T_2 - b_f) = x_f(T_2 - b_f),$$
$$|x_1(t) - x_f(t)| \leq \epsilon \text{ for all } t \in [T_1, T_1 + b_f],$$

$$|x_2(t) - x_f(t)| \leq \epsilon \text{ for all } t \in [T_2 - b_f, T_2],$$
$$I^f(T_1, T_1 + b_f, x_1, u_1) \leq I^f(T_1, T_1 + b_f, x_f, u_f) + \epsilon,$$
$$I^f(T_2 - b_f, T_2, x_2, u_2) \leq I^f(T_2 - b_f, T_2, x_f, u_f) + \epsilon. \tag{3.5}$$

By (3.2) and (3.5) there exists a trajectory-control pair

$$\tilde{x} : [T_1, T_2] \to R^n, \ \tilde{u} : [T_1, T_2] \to R^m$$

such that

$$\tilde{x}(t) = x_1(t), \ \tilde{u}(t) = u_1(t), \ t \in [T_1, T_1 + b_f],$$
$$\tilde{x}(t) = x_f(t), \ \tilde{u}(t) = u_f(t), \ t \in (T_1 + b_f, T_2 - b_f),$$
$$\tilde{x}(t) = x_2(t), \ \tilde{u}(t) = u_2(t), \ t \in [T_2 - b_f, T_2]. \tag{3.6}$$

By (3.5) and (3.6),

$$\tilde{x}(T_i) = \xi_i, \ i = 1, 2. \tag{3.7}$$

By (3.1), (3.2), (3.5)–(3.7),

$$U^f(T_1, T_2, \xi_1, \xi_2) \leq I^f(T_1, T_2, \tilde{x}, \tilde{u})$$
$$= I^f(T_1, T_1 + b_f, \tilde{x}, \tilde{u}) + I^f(T_1 + b_f, T_2 - b_f, \tilde{x}, \tilde{u}) + I^f(T_2 - b_f, T_2, \tilde{x}, \tilde{u})$$
$$= I^f(T_1, T_1 + b_f, x_1, u_1) + I^f(T_1 + b_f, T_2 - b_f, x_f, u_f) + I^f(T_2 - b_f, T_2, x_2, u_2)$$
$$\leq I^f(T_1, T_2, x_f, u_f) + 2\epsilon = I^f(T_1, T_2, x_f, u_f) + \epsilon_0.$$

Lemma 3.9 is proved. □

Lemma 3.10. *Let $\epsilon_0 > 0$. Then there exists a number $\delta > 0$ such that for each $(T_i, \xi_i) \in A$, $i = 1, 2$ which satisfy $T_2 > T_1$ and*

$$|\xi_i - x_f(T_i)| \leq \delta, \ i = 1, 2$$

the following relation holds:

$$U^f(T_1, T_2, \xi_1, \xi_2) \geq I^f(T_1, T_2, x_f, u_f) - \epsilon_0.$$

Proof. Let

$$\epsilon = \epsilon_0/8 \tag{3.8}$$

and let $\delta > 0$ be as guaranteed by B(iv). Let $T_1 \in R^1$, $T_2 > T_1$,

$$(T_i, \xi_i) \in A, \ i = 1, 2 \tag{3.9}$$

and

$$|\xi_i - x_f(T_i)| \le \delta, \ i = 1, 2. \tag{3.10}$$

We show that

$$U^f(T_1, T_2, \xi_1, \xi_2) \ge I^f(T_1, T_2, x_f, u_f) - \epsilon_0.$$

We may assume without loss of generality that

$$U^f(T_1, T_2, \xi_1, \xi_2) < \infty.$$

By (3.9), (3.10) and B(iv) which holds with ϵ and δ there exist trajectory-control pairs

$$x_1 : [T_1 - b_f, T_1] \to R^n, \ u_1 : [T_1 - b_f, T_1] \to R^m,$$
$$x_2 : [T_2, T_2 + b_f] \to R^n, \ u_2 : [T_2, T_2 + b_f] \to R^m$$

such that

$$x_1(T_1 - b_f) = x_f(T_1 - b_f),$$
$$x_1(T_1) = \xi_1,$$
$$x_2(T_2) = \xi_2,$$
$$x_2(T_2 + b_f) = x_f(T_2 + b_f),$$
$$|x_1(t) - x_f(t)| \le \epsilon \text{ for all } t \in [T_1 - b_f, T_1],$$
$$|x_2(t) - x_f(t)| \le \epsilon \text{ for all } t \in [T_2, T_2 + b_f],$$
$$I^f(T_1 - b_f, T_1, x_1, u_1) \le I^f(T_1 - b_f, T_1, x_f, u_f) + \epsilon,$$
$$I^f(T_2, T_2 + b_f, x_2, u_2) \le I^f(T_2, T_2 + b_f, x_f, u_f) + \epsilon. \tag{3.11}$$

By (3.11) there exists a trajectory-control pair

$$x : [T_1 - b_f, T_2 + b_f] \to R^n, \ u : [T_1 - b_f, T_2 + b_f] \to R^m$$

such that

$$x(t) = x_1(t), \ u(t) = u_1(t), \ t \in [T_1 - b_f, T_1],$$
$$x(t) = x_2(t), \ u(t) = u_2(t), \ t \in [T_2, T_2 + b_f], \tag{3.12}$$
$$I^f(T_1, T_2, x, u) \le U^f(T_1, T_2, \xi_1, \xi_2) + \epsilon/4. \tag{3.13}$$

By (3.11) and (3.12),

$$x(T_1 - b_f) = x_f(T_1 - b_f),$$
$$x(T_2 + b_f) = x_f(T_2 + b_f). \tag{3.14}$$

By (3.1), (3.12), (3.14), and B(i),

$$I^f(T_1 - b_f, T_2 + b_f, x_f, u_f) \le I^f(T_1 - b_f, T_2 + b_f, x, u)$$
$$= I^f(T_1 - b_f, T_1, x_1, u_1) + I^f(T_1, T_2, x, u) + I^f(T_2, T_2 + b_f, x_2, u_2)$$
$$\le I^f(T_1 - b_f, T_1, x_f, u_f) + \epsilon + I^f(T_1, T_2, x, u) + I^f(T_2, T_2 + b_f, x_f, u_f) + \epsilon$$

and

$$I^f(T_1, T_2, x_f, u_f) \le I^f(T_1, T_2, x, u) + 2\epsilon. \tag{3.15}$$

By (3.8), (3.13), and (3.15),

$$I^f(T_1, T_2, x_f, u_f) \le U^f(T_1, T_2, \xi_1, \xi_2) + 3\epsilon$$
$$\le U^f(T_1, T_2, \xi_1, \xi_2) + \epsilon_0.$$

Lemma 3.10 is proved. $\qquad\qquad\qquad\qquad\qquad\qquad\qquad\qquad\qquad\qquad\qquad$ □

Lemmas 3.9 and 3.10 imply the following result.

Lemma 3.11. *Let $\epsilon > 0$. Then there exists a number $\delta > 0$ such that for each $(T_i, \xi_i) \in A$, $i = 1, 2$ which satisfy*

$$T_2 \ge T_1 + 2b_f$$

and

$$|\xi_i - x_f(T_i)| \le \delta, \ i = 1, 2$$

the following relation holds:

$$|U^f(T_1, T_2, \xi_1, \xi_2) - I^f(T_1, T_2, x_f, u_f)| \le \epsilon.$$

3.3 Proof of Proposition 3.3

There exists a number $S > 0$ such that

$$I^f(T_1, T_2, x, u) \le \inf\{U^f(T_1, T_2, y_1, y_2) : (T_i, y_i) \in A, \ i = 1, 2\} + S \tag{3.16}$$

for each $T_1 \ge \tau$ and each $T_2 > T_1$.

Let a bounded continuous function $X_f : R^1 \to R^n$ be as guaranteed by the turnpike property. It follows from Remark 3.1 and the turnpike property that

$$x_f(t) = X_f(t) \text{ for all } t \in R^1. \tag{3.17}$$

By (3.16) and (LSC) property, for each integer $N \geq 1$ there exists a trajectory-control pair

$$x_N : [\tau, \tau + N] \to R^n, \ u_N : [\tau, \tau + N] \to R^m$$

such that

$$x_N(\tau) = x(\tau),$$

$$I^f(\tau, \tau + N, x_N, u_N) = \sigma^f(\tau, \tau + N, x(\tau)). \tag{3.18}$$

Let $k < N$ be integers. Relations (3.16) and (3.18) imply that

$$
\begin{aligned}
I^f(\tau, \tau + k, x_N, u_N) &= I^f(\tau, \tau + N, x_N, u_N) - I^f(\tau + k, \tau + N, x_N, u_N) \\
&\leq I^f(\tau, \tau + N, x, u) - I^f(\tau + k, \tau + N, x, u) + S \\
&= I^f(\tau, \tau + k, x, u) + S.
\end{aligned} \tag{3.19}
$$

Therefore for each integer $k \geq 0$ the sequence

$$\{I^f(\tau + k, \tau + k + 1, x_N, u_N)\}_{N=k+1}^{\infty}$$

is bounded. By (LSC) property there exists a subsequence $\{(x_{N_k}, u_{N_k})\}_{k=1}^{\infty}$ and a trajectory-control pair

$$x_* : [\tau, \infty) \to R^n, \ u_* : [\tau, \infty) \to R^m$$

such that

$$I^f(\tau + j, \tau + j + 1, x_*, u_*) \leq \liminf_{k \to \infty} I^f(\tau + j, \tau + j + 1, x_{N_k}, u_{N_k}) \tag{3.20}$$

for each integer $j \geq 1$ and

$$x_{N_k}(t) \to x_*(t) \text{ as } k \to \infty \text{ for any } t \in [\tau, \infty). \tag{3.21}$$

Clearly,

$$x_*(\tau) = x(\tau). \tag{3.22}$$

We show that

$$x_* : [\tau, \infty) \to R^n, \ u_* : [\tau, \infty) \to R^m$$

is an (f)-good trajectory-control pair. It follows from (3.20), and (3.19) that for each integer $j \geq 1$

$$I^f(\tau, \tau + j, x_*, u_*) \leq I^f(\tau, \tau + j, x, u) + S. \tag{3.23}$$

Let

$$T_1 \geq \tau \text{ and } T_2 > T_1.$$

Fix an integer $q > T_2 - \tau$. By (3.23),

$$I^f(T_1, T_2, x_*, u_*)$$
$$= I^f(\tau, \tau + q, x_*, u_*) - I^f(\tau, T_1, x_*, u_*) - I^f(T_2, \tau + q, x_*, u_*)$$
$$\leq I^f(\tau, \tau + q, x, u) + S - I^f(\tau, T_1, x, u) + S$$
$$- I^f(T_2, \tau + q, x, u) + S = I^f(T_1, T_2, x, u) + 3S.$$

Together with (3.16) this implies that (x_*, u_*) is an (f)-good trajectory-control pair.
  We show that (x_*, u_*) is an (f)-minimal trajectory-control pair. Assume the contrary. Then there exists an integer $Q \geq 1$ such that

$$I^f(\tau, \tau + Q, x_*, u_*) - U^f(\tau, \tau + Q, x_*(\tau), x_*(\tau + Q)) > 0. \tag{3.24}$$

Set

$$\Delta = 4^{-1}[I^f(\tau, \tau + Q, x_*, u_*) - U^f(\tau, \tau + Q, x_*(\tau), x_*(\tau + Q))]. \tag{3.25}$$

There exists a trajectory-control pair

$$y : [\tau, \tau + Q] \to R^n, \ v : [\tau, \tau + Q] \to R^m$$

such that

$$y(\tau) = x_*(\tau),$$
$$y(\tau + Q) = x_*(\tau + Q),$$
$$I^f(\tau, \tau + Q, x_*, u_*) - I^f(\tau, \tau + Q, y, v) \geq 2\Delta. \tag{3.26}$$

There exists $\delta \in (0, \Delta)$ such that Lemma 3.11 holds with $\epsilon = 32^{-1}\Delta(b_f + 1)^{-1}$. Since (x_*, u_*) is an (f)-good trajectory-control pair, it follows from (3.17) and Proposition 3.2 that

$$x_*(t) - x_f(t) \to 0 \text{ as } t \to \infty.$$

Therefore there exists an integer $T_0 \geq 1$ such that

$$|x_*(t) - x_f(t)| \leq 16^{-1}\delta \text{ for all } t \geq \tau + T_0. \tag{3.27}$$

Fix an integer

$$j \geq 4 + 4b_f. \tag{3.28}$$

By (3.20) and (3.21) there exists an integer k for which

$$k \geq Q + j + T_0,$$

$$I^f(\tau, \tau + T_0 + Q, x_*, u_*) \leq I^f(\tau, \tau + T_0 + Q, x_{N_k}, u_{N_k}) + 8^{-1}\delta,$$

$$|x_*(\tau + i) - x_{N_k}(\tau + i)| < 8^{-1}\delta, \ i = T_0 + Q, \ T_0 + Q + j. \tag{3.29}$$

By (3.26)–(3.29), the choice of δ and Lemma 3.11, there exists a trajectory-control pair

$$\tilde{x} : [\tau, \tau + N_k] \to R^n, \ \tilde{u} : [\tau, \tau + N_k] \to R^m$$

such that

$$\tilde{x}(t) = y(t), \ \tilde{u}(t) = v(t) \text{ for all } t \in [\tau, \tau + Q],$$

$$\tilde{x}(t) = x_*(t), \ \tilde{u}(t) = u_*(t) \text{ for all } t \in (\tau + Q, \tau + T_0 + Q],$$

$$\tilde{x}(t) = x_{N_k}(t), \ \tilde{u}(t) = u_{N_k}(t) \text{ for all } t \in [\tau + Q + T_0 + j, \tau + N_k],$$

$$I^f(\tau + T_0 + Q, \tau + Q + T_0 + j, \tilde{x}, \tilde{u})$$

$$= U^f(\tau + T_0 + Q, \tau + Q + T_0 + j, \tilde{x}(\tau + T_0 + Q), \tilde{x}(\tau + Q + T_0 + j)). \tag{3.30}$$

Relations (3.18), (3.30), (3.26), and (3.22) imply that

$$I^f(\tau, \tau + N_k, \tilde{x}, \tilde{u}) \geq I^f(\tau, \tau + N_k, x_{N_k}, u_{N_k}). \tag{3.31}$$

It follows from (3.27)–(3.30), (3.18) B(i), the choice of δ and Lemma 3.11 that

$$|I^f(\tau + T_0 + Q, \tau + Q + T_0 + j, \tilde{x}, \tilde{u})$$
$$-I^f(\tau + T_0 + Q, \tau + Q + T_0 + j, x_f, u_f)| \le 32^{-1}\Delta,$$
$$|I^f(\tau + T_0 + Q, \tau + Q + T_0 + j, x_{N_k}, u_{N_k})$$
$$-I^f(\tau + T_0 + Q, \tau + Q + T_0 + j, x_f, u_f)| \le 32^{-1}\Delta. \tag{3.32}$$

By (3.30), (3.26), (3.29), and (3.32),

$$I^f(\tau, \tau + N_k, \tilde{x}, \tilde{u}) - I^f(\tau, \tau + N_k, x_{N_k}, u_{N_k})$$
$$= I^f(\tau, \tau + Q, y, v) - I^f(\tau, \tau + Q, x_{N_k}, u_{N_k})$$
$$+I^f(\tau + Q, \tau + Q + T_0, x_*, u_*) - I^f(\tau + Q, \tau + Q + T_0, x_{N_k}, u_{N_k})$$
$$+I^f(\tau + T_0 + Q, \tau + Q + T_0 + j, \tilde{x}, \tilde{u})$$
$$-I^f(\tau + T_0 + Q, \tau + Q + T_0 + j, x_{N_k}, u_{N_k})$$
$$\le -2\Delta + I^f(\tau, \tau + Q + T_0, x_*, u_*) - I^f(\tau, \tau + Q + T_0, x_{N_k}, u_{N_k}) + 16^{-1}\Delta + \delta$$
$$\le -2\Delta + 16^{-1}\Delta + 8^{-1}\delta \le -\Delta.$$

This contradicts (2.31). The obtained contradiction we have reached proves that (x_*, u_*) is an (f)-minimal trajectory-control pair. Proposition 3.3 is proved.

3.4 Proof of Proposition 3.4

Let a bounded continuous function $X_f : R^1 \to R^n$ be as guaranteed by the turnpike property. It follows from Remark 3.1, assumption (B), and the turnpike property that

$$x_f(t) = X_f(t) \text{ for all } t \in R^1. \tag{3.33}$$

By Proposition 3.2

$$x(t) - X_f(t) \to 0 \text{ as } t \to \infty. \tag{3.34}$$

Let

$$y : [\tau, \infty) \to R^n, \ v : [\tau, \infty) \to R^m$$

be a trajectory-control pair satisfying

$$y(\tau) = x(\tau).$$

We will show that

$$\limsup_{T\to\infty}[I^f(\tau, T, x, u) - I^f(\tau, T, y, v)] \le 0. \qquad (3.35)$$

Assume the contrary. Then there exists a number $\epsilon > 0$ for which

$$\limsup_{T\to\infty}[I^f(\tau, T, x, u) - I^f(\tau, T, y, v)] > 2\epsilon. \qquad (3.36)$$

By Theorem 2.1 and Remark 3.1, we may assume without loss of generality that (y, v) is an (f)-good trajectory-control pair. Proposition 3.2 and (3.33) imply that

$$y(t) - x_f(t) \to 0 \text{ as } t \to \infty. \qquad (3.37)$$

Set

$$\epsilon_0 = 4^{-1}\epsilon. \qquad (3.38)$$

There exists $\delta \in (0, \epsilon_0)$ such that Lemma 3.11 holds with $\epsilon = \epsilon_0$ and δ. By (3.33), (3.34), and (3.37) there exists an integer $Q \ge 1$ such that

$$|x(t) - x_f(t)|, \ |y(t) - x_f(t)| \le 8^{-1}\delta, \ t \in [Q, \infty). \qquad (3.39)$$

It follows from (3.36) that there exists a number T_0 for which

$$T_0 > Q + 4,$$

$$I^f(\tau, T_0, x, u) - I^f(\tau, T_0, y, v) > \epsilon. \qquad (3.40)$$

Fix an integer $j \ge 2b_f + 8$. By (3.39) and (3.40), the choice of δ, (LSC) property and Lemma 3.11, there exists a trajectory-control pair

$$\tilde{x} : [\tau, \infty) \to R^n, \ \tilde{u} : [\tau, \infty) \to R^m$$

such that

$$\tilde{x}(t) = y(t), \ \tilde{u}(t) = v(t) \text{ for all } t \in [\tau, T_0],$$

$$\tilde{x}(t) = x(t), \ \tilde{u}(t) = u(t) \text{ for all } t \in [T_0 + j, \infty),$$

$$I^f(T_0, T_0 + j, \tilde{x}, \tilde{u}) = U^f(T_0, T_0 + j, \tilde{x}(T_0), \tilde{x}(T_0 + j)). \qquad (3.41)$$

It follows from (3.39)–(3.41), the proposition assumptions, assumption (B) the choice of δ and Lemma 3.11 that

$$|I^f(T_0, T_0 + j, \tilde{x}, \tilde{u}) - I^f(T_0, T_0 + j, x, u)| \le 2\epsilon_0. \qquad (3.42)$$

Since (x, u) is an (f)-minimal trajectory-control pair (3.41) implies that

$$I^f(\tau, T_0 + j, x, u) \leq I^f(\tau, T_0 + j, \tilde{x}, \tilde{u}). \tag{3.43}$$

On the other hand it follows from (3.41), (3.42), (3.40), and (3.38) that

$$I^f(\tau, T_0 + j, \tilde{x}, \tilde{u}) - I^f(\tau, T_0 + j, x, u)$$
$$\leq I^f(\tau, T_0, y, v) - I^f(\tau, T_0, x, u) + 2\epsilon_0$$
$$\leq -\epsilon + 2\epsilon_0 \leq -\epsilon_0.$$

This contradicts (3.43). The contradiction we have reached proves Proposition 3.4.

3.5   Proof of Theorem 3.6

Let a bounded continuous function $X_f : R^1 \to R^n$ be as guaranteed by the turnpike property. It follows from Remark 3.1, assumption (B), and the turnpike property that

$$x_f(t) = X_f(t) \text{ for all } t \in R^1. \tag{3.44}$$

Assume that (x, u) is an (f)-agreeable trajectory-control pair. We show that (x, u) is (f)-minimal.

Assume the contrary. Then there exists $S_0 > \tau$ such that

$$I^f(\tau, S_0, x, u) > U^f(\tau, S_0, x(\tau), x(S_0)).$$

This inequality implies that there exist a constant $\Delta_0 > 0$ and a trajectory-control pair

$$x_1 : [\tau, S_0] \to R^n, \ u_1 : [\tau, S_0] \to R^m$$

such that

$$x_1(\tau) = x(\tau), \ x_1(S_0) = x(S_0), \tag{3.45}$$

$$\infty > I^f(\tau, S_0, x, u) > I^f(\tau, S_0, x_1, u_1) + \Delta_0. \tag{3.46}$$

Since the trajectory-control pair (x, u) is (f)-agreeable there is $T > S_0$ and a trajectory-control pair

$$y : [\tau, T] \to R^n, \ v : [\tau, T] \to R^m$$

such that

$$y(t) = x(t), \ u(t) = v(t) \text{ for all } t \in [\tau, S_0], \tag{3.47}$$

$$I^f(\tau, T, y, v) = \sigma^f(\tau, T, x(\tau)) + \Delta_0/2 < \infty. \tag{3.48}$$

By (3.45) and (3.47) there exists a trajectory-control pair

$$y_1 : [\tau, T] \to R^n, \ v_1 : [\tau, T] \to R^m$$

such that

$$y_1(t) = x_1(t), \ v_1(t) = u_1(t) \text{ for all } t \in [\tau, S_0],$$

$$y_1(t) = y(t), \ v_1(t) = v(t) \text{ for all } t \in (S_0, T]. \tag{3.49}$$

By (3.49), (3.46)–(3.48),

$$\begin{aligned}
I^f(\tau, T, y_1, v_1) &= I^f(\tau, S_0, y_1, v_1) + I^f(S_0, T, y_1, v_1) \\
&= I^f(\tau, S_0, x_1, u_1) + I^f(S_0, T, y, v) \\
&< I^f(\tau, S_0, x, u) - \Delta_0 + I^f(S_0, T, y, v) \\
&= I^f(\tau, T, y, v) - \Delta_0 \le \sigma^f(\tau, T, x(\tau)) - \Delta_0/2.
\end{aligned}$$

This contradicts (3.49) and (3.45). The contradiction we have reached proves that (x, u) is an (f)-minimal trajectory-control pair.

Now assume that the trajectory-control pair (x, u) is (f)-minimal. We claim that (x, u) is an (f)-agreeable trajectory-control pair. By Remark 3.1 (x_f, u_f) is an (f)-good trajectory-control pair. Therefore there is $Q_1 > 0$ such that for each $T_1 \in R^1$ and each $T_2 > T_1$,

$$I^f(T_1, T_2, x_f, u_f) \le \inf\{U^f(T_1, T_2, \xi_1, \xi_2) : (T_i, \xi_i) \in A, \ i = 1, 2\} + Q_1. \tag{3.50}$$

Since the trajectory-control pair (x, u) is (f)-good there is $Q_2 > 0$ such that for each $T_1 \ge \tau$ and each $T_2 > T_1$,

$$I^f(T_1, T_2, x, u) \le \inf\{U^f(T_1, T_2, \xi_1, \xi_2) : (T_i, \xi_i) \in A, \ i = 1, 2\} + Q_2. \tag{3.51}$$

Let

$$T_0 > \tau \text{ and } \epsilon > 0.$$

By Lemma 3.11 there exists a number $\delta \in (0, 1)$ such that for each $(T_i, \xi_i) \in A$, $i = 1, 2$ which satisfy

$$T_2 \ge T_1 + 2b_f$$

and

$$|\xi_i - x_f(T_i)| \le \delta, \ i = 1, 2 \tag{3.52}$$

the following relation holds:

$$|U^f(T_1, T_2, \xi_1, \xi_2) - I^f(T_1, T_2, x_f, u_f)| \le \epsilon/8. \tag{3.53}$$

By Proposition 3.2 and (3.44) there exists $S_0 > T_0$ such that

$$|x(t) - x_f(t)| \le \delta \text{ for all } t \ge S_0. \tag{3.54}$$

It follows from the turnpike property and (3.44) that there exist

$$\Delta_0 > 0, \ \delta_0 \in (0, \min\{\delta, \ \epsilon/8\}) \tag{3.55}$$

such that for each pair of numbers

$$S_1 \in R^1, \ S_2 \ge S_1 + 2\Delta_0$$

and each trajectory-control pair

$$y : [S_1, S_2] \to R^n, \ v : [S_1, S_2] \to R^m$$

which satisfies

$$I^f(S_1, S_2, y, v)$$
$$\le \inf\{U^f(S_1, S_2, \xi_1, \xi_2) : \ (S_i, \xi_i) \in A, \ i = 1, 2\} + Q_1 + Q_2 + 4 \tag{3.56}$$

and

$$I^f(S_1, S_2, y, v) \le U^f(S_1, S_2, y(S_1), y(S_2)) + \delta_0 \tag{3.57}$$

the following inequality holds:

$$|y(t) - x_f(t)| \le \delta \text{ for all } t \in [S_1 + \Delta_0, S_2 - \Delta_0]. \tag{3.58}$$

Choose a number

$$T_\epsilon > T_0 + 4|S_0| + 2\Delta_0 + 2b_f. \tag{3.59}$$

Let $T \ge T_\epsilon$. There exists a trajectory-control pair

$$y_1 : [\tau, T] \to R^n, \ v_1 : [\tau, T] \to R^m$$

such that

$$y_1(\tau) = x(\tau), \tag{3.60}$$

$$I^f(\tau, T, y_1, v_1) \leq \sigma^f(\tau, T, x(\tau)) + \delta_0 < \infty. \tag{3.61}$$

By (3.60), (3.61), and (3.51),

$$I^f(\tau, T, y_1, v_1) \leq I^f(\tau, T, x, u) + 1$$
$$\leq \inf\{U^f(\tau, T, \xi_1, \xi_2) : (\tau, \xi_1), (T, \xi_2) \in A\} + Q_2 + 1. \tag{3.62}$$

By (3.61), (3.62), (3.59), and the choice of Δ_0, δ_0 [see (3.55), (3.57) and (3.56)],

$$|y_1(t) - x_f(t)| \leq \delta \text{ for all } t \in [\tau + \Delta_0, T - \Delta_0]. \tag{3.63}$$

By (3.59),

$$T - \Delta_0 - 2b_f \geq S_0. \tag{3.64}$$

By (3.64) and (3.54),

$$|x(T - \Delta_0 - 2b_f) - x_f(T - \Delta_0 - 2b_f)| \leq \delta, \tag{3.65}$$

$$|x(T - \Delta_0) - x_f(T - \Delta_0)| \leq \delta. \tag{3.66}$$

By (3.59),

$$\tau + \Delta_0 < \Delta_0 + T_0 < T - \Delta_0 - 2b_f < T - \Delta_0. \tag{3.67}$$

By (3.67) and (3.63),

$$|y_1(T - \Delta_0 - 2b_f) - x_f(T - \Delta_0 - 2b_f)| \leq \delta, \tag{3.68}$$

$$|y_1(T - \Delta_0) - x_f(T - \Delta_0)| \leq \delta. \tag{3.69}$$

By (3.65), (3.66), (3.68), (3.69), and the choice of δ [see (3.52) and (3.53)],

$$|I^f(T - \Delta_0 - 2b_f, T - \Delta_0, x_f, u_f)$$
$$-U^f(T - \Delta_0 - 2b_f, T - \Delta_0, \eta_1, \eta_2)| \leq \epsilon/8 \tag{3.70}$$

for all

$$\eta_1 \in \{x(T - \Delta_0 - 2b_f),\ y_1(T - \Delta_0 - 2b_f)\}$$

and all

$$\eta_2 \in \{x(T - \Delta_0),\ y_1(T - \Delta_0)\}.$$

There exists a trajectory-control pair

$$y_2 : [T - \Delta_0 - 2b_f, T - \Delta_0] \to R^n,\ v_2 : [T - \Delta_0 - 2b_f, T - \Delta_0] \to R^m$$

such that

$$
\begin{aligned}
y_2(T - \Delta_0 - 2b_f) &= y_1(T - \Delta_0 - 2b_f), \\
y_2(T - \Delta_0) &= x(T - \Delta_0), \qquad\qquad\qquad (3.71) \\
& I^f(T - \Delta_0 - 2b_f, T - \Delta_0, y_2, v_2)
\end{aligned}
$$

$$\le U^f(T - \Delta_0 - 2b_f, T - \Delta_0, y_2(T - \Delta_0 - 2b_f), y_2(T - \Delta_0)) + \epsilon/16. \quad (3.72)$$

By (3.71) and (3.60), the (f)-minimality of (x, u), (3.67), (3.72) and (3.70),

$$I^f(\tau, T - \Delta_0, x, u)$$

$$\le I^f(\tau, T - \Delta_0 - 2b_f, y_1, v_1) + I^f(T - \Delta_0 - 2b_f, T - \Delta_0, y_2, v_2)$$

$$\le I^f(\tau, T - \Delta_0 - 2b_f, y_1, v_1)$$

$$+ U^f(T - \Delta_0 - 2b_f, T - \Delta_0, y_2(T - \Delta_0 - 2b_f), y_2(T - \Delta_0)) + \epsilon/16$$

$$\le I^f(\tau, T - \Delta_0 - 2b_f, y_1, v_1)$$

$$+ I^f(T - \Delta_0 - 2b_f, T - \Delta_0, x_f, u_f) + \epsilon/16 + \epsilon/16. \qquad (3.73)$$

By (3.67) and (3.70),

$$I^f(T - \Delta_0 - 2b_f, T - \Delta_0, y_1, v_1)$$

$$\ge U^f(T - \Delta_0 - 2b_f, T - \Delta_0, y_1(T - \Delta_0 - 2b_f), y_1(T - \Delta_0))$$

$$\ge I^f(T - \Delta_0 - 2b_f, T - \Delta_0, x_f, u_f) - \epsilon/8. \qquad (3.74)$$

By (3.73) and (3.74),

$$I^f(\tau, T - \Delta_0, x, u) \le I^f(\tau, T - \Delta_0, y_1, v_1) + \epsilon/4 + \epsilon/16. \qquad (3.75)$$

By (3.70) there exists a trajectory-control pair

$$y : [\tau, T] \to R^n, \ v : [\tau, T] \to R^m$$

such that

$$y(t) = x(t), \ v(t) = u(t), \ t \in [\tau, T - \Delta_0 - 2b_f],$$

$$y(t) = y_1(t), \ v(t) = v_1(t), \ t \in [T - \Delta_0, T],$$

$$I^f(T - \Delta_0 - 2b_f, T - \Delta_0, y, v)$$

$$\leq U^f(T - \Delta_0 - 2b_f, T - \Delta_0, y(T - \Delta_0 - 2b_f), y(T - \Delta_0)) + \epsilon/16. \quad (3.76)$$

By (3.76) and (3.67),

$$y(t) = x(t) \text{ for all } t \in [0, T_0]. \tag{3.77}$$

By (3.76), (3.70), and (3.75) and the (f)-minimality of (x, u),

$$I^f(\tau, T, y, v) = I^f(\tau, T - \Delta_0 - 2b_f, x, u)$$

$$+ I^f(T - \Delta_0, T, y_1, v_1) + I^f(T - \Delta_0 - 2b_f, T - \Delta_0, x_f, u_f) + \epsilon/8 + \epsilon/8$$

$$\leq I^f(T - \Delta_0, T, y_1, v_1) + I^f(\tau, T - \Delta_0 - 2b_f, x, u)$$

$$+ U^f(T - \Delta_0 - 2b_f, T - \Delta_0, x(T - \Delta_0 - 2b_f), x(T - \Delta_0)) + 3(\epsilon/8)$$

$$= I^f(T - \Delta_0, T, y_1, v_1) + I^f(\tau, T - \Delta_0 - 2b_f, x, u)$$

$$+ I^f(T - \Delta_0 - 2b_f, T - \Delta_0, x, u) + 3\epsilon/8$$

$$= I^f(T - \Delta_0, T, y_1, v_1) + I^f(\tau, T - \Delta_0, x, u) + 3\epsilon/8$$

$$\leq I^f(T - \Delta_0, T, y_1, v_1) + I^f(\tau, T - \Delta_0, y_1, v_1) + 5\epsilon/8 + \epsilon/16. \quad (3.78)$$

In view of (3.78), (3.61), and (3.55),

$$I^f(\tau, T, y, v) \leq \sigma^f(\tau, T, x(\tau)) + \epsilon. \tag{3.79}$$

Thus for any $T \geq T_\epsilon$ there is a trajectory-control pair

$$y : [\tau, T] \to R^n, \ v : [\tau, T] \to R^m$$

satisfying (3.77) and (3.79). Therefore (x, u) is an (f)-agreeable trajectory-control pair. Theorem 3.6 is proved.

3.6 Proof of Theorem 3.7

Let $f \in \mathfrak{M}_{reg}$ possess the turnpike property with the turnpike X_f, $M > 0$ and $\epsilon > 0$. By Remark 3.1,

$$x_f(t) = X_f(t) \text{ for all } t \in R^1. \tag{3.80}$$

Set

$$M_0 = M + 4 + \sup\{|I^f(t, t + 2b_f, x_f, u_f)| : t \in R^1\} + 2a_0 b_f. \tag{3.81}$$

Lemma 3.12. *Let* $\tau \in R^1$,

$$x : [\tau, \infty) \to R^n, \ u : [\tau, \infty) \to R^m$$

be an (f)*-good and* (f)*-minimal trajectory-control pair such that* $x(\tau)$ *is an* (f, τ, M)*-good point. Then there exists* $\tilde{T} > \tau$ *such that for each* $T \geq \tilde{T}$,

$$I^f(\tau, T, x, u)$$

$$\leq \inf\{U^f(\tau, T, \xi_1, \xi_2) : (\tau, \xi_1), (T, \xi_2) \in A\} + M_0.$$

Proof. There exists an (f)-good trajectory-control pair

$$y : [\tau, \infty) \to R^n, \ v : [\tau, \infty) \to R^m$$

such that

$$y(\tau) = x(\tau),$$

$$I^f(S_1, S_2, y, v)$$

$$\leq \inf\{U^f(S_1, S_2, \xi_1, \xi_2) : (S_i, \xi_i) \in A, \ i = 1, 2\} + M \tag{3.82}$$

for each $S_1 \geq \tau$ and each $S_2 > S_1$. By Proposition 3.2, (3.80), and (3.82),

$$\lim_{t \to \infty} |x(t) - x_f(t)| = 0, \tag{3.83}$$

$$\lim_{t \to \infty} |y(t) - x_f(t)| = 0. \tag{3.84}$$

There exists $\delta \in (0, 1)$ such that B(iv) holds with $\epsilon = 1$. By (3.83) and (3.84) there exists

$$\tilde{T} > \tau + 2b_f + 4 \tag{3.85}$$

such that

$$|x(t) - x_f(t)| \le \delta \text{ for all } t \ge \tilde{T} - 2b_f, \tag{3.86}$$

$$|y(t) - x_f(t)| \le \delta \text{ for all } t \ge \tilde{T} - 2b_f, \tag{3.87}$$

Let

$$T \ge \tilde{T}. \tag{3.88}$$

By (3.88), (3.87), (3.86), the choice of δ and B(iv) (which holds with $\epsilon = 1$) there exist trajectory-control pairs

$$y_1 : [T - 2b_f, T - b_f] \to R^n, \ v_1 : [T - 2b_f, T - b_f] \to R^m,$$
$$y_2 : [T - b_f, T] \to R^n, \ v_2 : [T - b_f, T] \to R^m$$

such that

$$y_1(T - 2b_f) = y(T - 2b_f),$$
$$y_1(T - b_f) = x_f(T - b_f),$$
$$y_2(T - b_f) = x_f(T - b_f),$$
$$y_2(T) = x(T),$$
$$I^f(T - 2b_f, T - b_f, y_1, v_1) \le I^f(T - 2b_f, T - b_f, x_f, u_f) + 1,$$
$$I^f(T - b_f, T, y_2, v_2) \le I^f(T - b_f, T, x_f, u_f) + 1. \tag{3.89}$$

By (3.89) there exists a trajectory-control pair

$$\tilde{y} : [\tau, T] \to R^n, \ \tilde{v} : [\tau, T] \to R^m$$

such that

$$\tilde{y}(t) = y(t), \ \tilde{v}(t) = v(t), \ t \in [\tau, T - 2b_f],$$
$$\tilde{y}(t) = y_1(t), \ \tilde{v}(t) = v_1(t), \ t \in (T - 2b_f, T - b_f],$$
$$\tilde{y}(t) = y_2(t), \ \tilde{v}(t) = v_2(t), \ t \in (T - b_f, T]. \tag{3.90}$$

By (3.90), (3.89), (3.81), and the (f)-minimality of (x, u),

$$I^f(\tau, T, x, u) \le I^f(\tau, T, \tilde{y}, \tilde{v}). \tag{3.91}$$

By (3.90) and (3.89),

$$I^f(\tau, T, \tilde{y}, \tilde{v}) = I^f(\tau, T - 2b_f, y, v)$$
$$+I^f(T - 2b_f, T - b_f, y_1, v_1) + I^f(T - b_f, T, y_2, v_2)$$
$$= I^f(\tau, T, y, v) - I^f(T - 2b_f, T, y, v)$$
$$+I^f(T - 2b_f, T - b_f, x_f, u_f) + 1 + I^f(T - b_f, T, x_f, u_f) + 1$$
$$\leq I^f(\tau, T, y, v) + 2a_0 b_f + I^f(T - 2b_f, T, x_f, y_f) + 2. \qquad (3.92)$$

In view of (3.91), (3.92), and (3.82),

$$I^f(\tau, T, x, u) \leq I^f(\tau, T, y, v) + 2 + 2a_0 b_f$$
$$+I^f(T - 2b_f, T, x_f, u_f)$$
$$\leq 2 + 2a_0 b_f + I^f(T - 2b_f, T, x_f, u_f)$$
$$+ \inf\{U^f(\tau, T, \xi_1, \xi_2) : (\tau, \xi_1), (T, \xi_2) \in A\} + M$$
$$\leq \inf\{U^f(\tau, T, \xi_1, \xi_2) : (\tau, \xi_1), (T, \xi_2) \in A\} + M_0.$$

Lemma 3.12 is proved.   □

Proof of Theorem 3.7. It follows from the turnpike property and (3.80) that there exist numbers $\Delta > 0$, $\delta > 0$ such that for each pair of number

$$T_1 \in R^1, \ T_2 \geq T_1 + 2\Delta$$

and each trajectory-control pair

$$x : [T_1, T_2] \to R^n, \ u : [T_1, T_2] \to R^m$$

which satisfies

$$I^f(T_1, T_2, x, u) \leq \inf\{U^f(T_1, T_2, \xi_1, \xi_2) : (T_i, \xi_i) \in A, \ i = 1, 2\} + M_0 \quad (3.93)$$

and

$$I^f(T_1, T_2, x, u) \leq U^f(T_1, T_2, x(T_1), x(T_2)) + \delta \qquad (3.94)$$

the following inequality holds:

$$|x(t) - x_f(t)| \leq \epsilon \text{ for all } t \in [T_1 + \Delta, T_2 - \Delta].$$

Let $\tau \in R^1$,

$$x : [\tau, \infty) \to R^n, \ u : [\tau, \infty) \to R^m$$

be an (f)-good and (f)-minimal trajectory-control pair such that $x(\tau)$ is an (f, τ, M)-good point. By Lemma 3.12 there exists $\tilde{T} > \tau$ such that for each $T \geq \tilde{T}$,

$$I^f(\tau, T, x, u)$$

$$\leq \inf\{U^f(\tau, T, \xi_1, \xi_2) : \ (\tau, \xi_1), \ (T, \xi_2) \in A\} + M_0. \tag{3.95}$$

Let

$$T \geq \tilde{T} + 2\Delta.$$

Then (3.95) holds and since (x, u) is an (f)-minimal trajectory-control pair

$$I^f(\tau, T, x, u) = U^f(\tau, T, x(\tau), x(T)). \tag{3.96}$$

By (3.95), (3.96), and the choice of Δ, δ [see (3.93) and (3.94)]

$$|x(t) - x_f(t)| \leq \epsilon \text{ for all } t \in [\tau + \Delta, T - \Delta].$$

Since the inequality above holds for each $T \geq \tilde{T} + 2\Delta$, we conclude that

$$|x(t) - x_f(t)| \leq \epsilon \text{ for all } t \geq \tau + \Delta.$$

Theorem 3.7 is proved.   □

3.7 Proof of Theorem 3.8

We suppose that sum over empty set is zero. Since $f \in \mathfrak{M}_{reg}$ possesses the turnpike property (with the turnpike X_f) it follows from Remark 3.1 and assumption B(i) that

$$x_f(t) = X_f(t) \text{ for all } t \in R^1. \tag{3.97}$$

By Remark 3.1 (x_f, u_f) is an (f)-good trajectory-control pair. Therefore there is $Q_0 > 0$ such that for each $T_1 \in R^1$ and each $T_2 > T_1$,

$$I^f(T_1, T_2, x_f, u_f) \leq \inf\{U^f(T_1, T_2, \xi_1, \xi_2) : \ (T_i, \xi_i) \in A, \ i = 1, 2\} + Q_0. \tag{3.98}$$

Lemma 3.13. *Let* $-\infty < T_1 \le S_1 < S_2 \le T_2 < \infty$ *and*

$$x : [T_1, T_2] \to R^n, \ u : [T_1, T_2] \to R^m$$

be a trajectory-control pair such that

$$I^f(T_1, T_2, x, u) \le \inf\{U^f(T_1, T_2, \xi_1, \xi_2) : (T_i, \xi_i) \in A, \ i = 1, 2\} + M. \quad (3.99)$$

Then

$$I^f(S_1, S_2, x, u) \le \inf\{U^f(S_1, S_2, \xi_1, \xi_2) : (S_i, \xi_i) \in A, \ i = 1, 2\} + M + 3Q_0.$$

Proof. By (3.99) and (3.98),

$$I^f(S_1, S_2, x, u)$$
$$= I^f(T_1, T_2, x, u) - I^f(T_1, S_1, x, u) - I^f(S_2, T_2, x, u)$$
$$\le I^f(T_1, T_2, x_f, u_f) + M - (I^f(T_1, S_1, x_f, u_f) - Q_0) - (I^f(S_2, T_2, x_f, u_f) - Q_0)$$
$$= I^f(S_1, S_2, x_f, u_f) + M + 3Q_0$$
$$\le \inf\{U^f(S_1, S_2, \xi_1, \xi_2) : (S_i, \xi_i) \in A, \ i = 1, 2\} + M + 3Q_0.$$

Lemma 3.13 is proved. $\qquad\qquad\qquad\qquad\qquad\qquad\qquad\qquad\qquad\qquad\quad \square$

Proof of Theorem 3.8. By the turnpike property and (3.97) there exist numbers $\Delta > 0$, $\delta > 0$ such that for each pair of number

$$S_1 \in R^1, \ S_2 \ge S_1 + 2\Delta$$

and each trajectory-control pair

$$x : [S_1, S_2] \to R^n, \ u : [S_1, S_2] \to R^m$$

which satisfies

$$I^f(S_1, S_2, x, u) \le U^f(S_1, S_2, x(S_1), x(S_2)) + \delta \qquad\qquad (3.100)$$

and

$$I^f(S_1, S_2, x, u) \le \inf\{U^f(S_1, S_2, \xi_1, \xi_2) : (S_i, \xi_i) \in A, \ i = 1, 2\} + M + 3Q_0 \qquad (3.101)$$

the following inequality holds:

$$|x(t) - x_f(t)| \le \epsilon \text{ for all } t \in [S_1 + \Delta, S_2 - \Delta]. \qquad\qquad (3.102)$$

Fix

$$l > 1 + 2\Delta, \tag{3.103}$$

a natural number

$$Q > 3M\delta^{-1} + 4 \tag{3.104}$$

and

$$L > lQ. \tag{3.105}$$

Assume that

$$T_1 \in R^1, \ T_2 \geq T_1 + L$$

and that a trajectory-control pair

$$x : [T_1, T_2] \to R^n, \ u : [T_1, T_2] \to R^m$$

satisfies

$$I^f(T_1, T_2, x, u) \leq \inf\{U^f(T_1, T_2, \xi_1, \xi_2) : \ (T_i, \xi_i) \in A, \ i = 1, 2\} + M. \tag{3.106}$$

By (3.106) and Lemma 3.13 for all numbers S_1, S_2 satisfying

$$T_1 \leq S_1 < S_2 \leq T_2$$

we have

$$I^f(S_1, S_2, x, u) \leq \inf\{U^f(S_1, S_2, \xi_1, \xi_2) : \ (S_i, \xi_i) \in A, \ i = 1, 2\} + M + 3Q_0. \tag{3.107}$$

Set

$$t_0 = T_1. \tag{3.108}$$

Assume that an integer $k \geq 0$ and we have defined a strictly increasing sequence

$$\{t_i\}_{i=0}^k \subset [T_1, T_2]$$

such that $t_k < T_2$ and for each integer i satisfying $1 \leq i \leq k$,

$$I^f(t_{i-1}, t_i, x, u) > U^f(t_{i-1}, t_i, x(t_{i-1}), x(t_i)) + \delta \tag{3.109}$$

and there a number s_i such that

$$t_{i-1} < s_i < t_i, \; s_i > t_i - 1,$$

$$I^f(t_{i-1}, s_i, x, u) \leq U^f(t_{i-1}, s_i, x(t_{i-1}), x(s_i)) + \delta. \tag{3.110}$$

(Clearly, for $k = 0$ our assumption holds.)
   By (3.106) and (3.109),

$$M \geq I^f(T_1, T_2, x, u) - U^f(T_1, T_2, x(T_1), x(T_2))$$

$$\geq \sum \{ I^f(t_{i-1}, t_i, x, u) - U^f(t_{i-1}, t_i, x(t_{i-1}), x(t_i)) :$$

$$i \text{ is an integer and } 1 \leq i \leq k \} \geq \delta k$$

and

$$k \leq M \delta^{-1}. \tag{3.111}$$

Let us define t_{k+1}. If

$$I^f(t_k, T_2, x, u) \leq U^f(t_k, T_2, x(t_k), x(T_2)) + \delta,$$

then we set $t_{k+1} = T_2$ and our construction is completed.
   Assume that

$$I^f(t_k, T_2, x, u) > U^f(t_k, T_2, x(t_k), x(T_2)) + \delta. \tag{3.112}$$

Set

$$\tilde{t}_{k+1} = \inf\{\tau \in (t_k, T_2] : \; I^f(t_k, \tau, x, u) > U^f(t_k, \tau, x(t_k), x(\tau)) + \delta\}. \tag{3.113}$$

Clearly, $\tilde{t}_{k+1}$ is well defined and

$$T_2 \geq \tilde{t}_{k+1} > t_k.$$

If $\tilde{t}_{k+1} = T_2$, then we set $t_{k+1} = T_2$ and our construction is completed.
   If $\tilde{t}_{k+1} < T_2$, then there exist

$$t_{k+1} \in (\tilde{t}_{k+1}, T_2), \; s_{k+1} \in (t_k, \tilde{t}_{k+1})$$

such that

$$t_{k+1} - s_{k+1} < 1,$$

and in this case the assumption we made for k also holds for $k + 1$.
   By (3.111) our sequence $\{t_i\}$ is finite. Let $k \geq 0$ be an integer such that t_{k+1} is its last element. By (3.111),

$$k + 1 \leq M \delta^{-1} + 1. \tag{3.114}$$

It follows from the construction of the sequence that

$$t_{k+1} = T_2$$

and that for each integer $i \in \{1, \ldots, k+1\}$ there a number $s_i \in (t_{i-1}, t_i]$ such that

$$s_i \geq t_i - 1 \tag{3.115}$$

and (3.110) holds. Set

$$A = \{i \in \{1, \ldots, k+1\} : t_i - t_{i-1} \geq 2\Delta + 1\}. \tag{3.116}$$

Let

$$i \in A. \tag{3.117}$$

By (3.117), (3.116), and (3.115),

$$s_i - t_{i-1} \geq 2\Delta \tag{3.118}$$

and (3.110) holds. By (3.107),

$$I^f(t_{i-1}, s_i, x, u) \leq \inf\{U^f(t_{i-1}, s_i, \xi_1, \xi_2) : (t_{i-1}, \xi_1), (s_i, \xi_2) \in A\} + M + 3Q_0. \tag{3.119}$$

By (3.110), (3.119), (2.118), and the choice of Δ, δ [see (3.100) and (3.101)],

$$|x(t) - x_f(t)| \leq \epsilon \text{ for all } t \in [t_{i-1} + \Delta, s_i - \Delta] \tag{3.120}$$

for all $i \in A$.
   By (3.120),

$$\{t \in [T_1, T_2] : |x(t) - x_f(t)| > \epsilon\}$$
$$\subset \cup\{[t_{i-1}, t_i] : i \in \{1, \ldots, k+1\} \setminus A\} \cup \{[t_{i-1}, t_{i-1} + \Delta] : i \in A\}$$
$$\cup\{[s_i - \Delta, t_i] : i \in A\}. \tag{3.121}$$

Clearly, the right-hand side of (3.121) is a finite union of intervals, by (3.116), (3.115), (3.103) their maximal length does not exceed $2\Delta + 1 \leq l$ and in view of (3.104) and (3.114) their number does not exceed

$$3(k+1) \leq 3M\delta^{-1} + 3 \leq Q.$$

Theorem 3.8 is proved. $\square$

Chapter 4
Linear Control Systems

In this chapter we continue to use the notation and the definitions of Chaps. 2 and 3.

4.1 The Class of Problems

Consider the control system described by (2.1)–(2.5). Assume that

$$M = R^{n+m+1}, \ U(t,x) = R^m, \ (t,x) \in R^{n+1}, \tag{4.1}$$

$$G(t,x,u) = Ax + Bu, \ t \in R^1, \ x \in R^n, \ u \in R^m, \tag{4.2}$$

where A and B are given matrices of dimensions $n \times n$ and $n \times m$. We also assume that linear system

$$x'(t) = Ax(t) + Bu(t) \tag{4.3}$$

is controllable.

Denote by $\mathfrak{M}_c$ the set of all continuous functions $f \in \mathfrak{M}$ which satisfy the following assumptions:

D(i) for each $(t,x) \in R^{n+1}$ the function $f(t,x,\cdot) : R^n \to R^1$ is convex;

D(ii) for each $K > 0$ there exists a constant $a_{f,K} > 0$ and an increasing function

$$\psi_{f,K} : [0,\infty) \to [0,\infty)$$

such that

$$\psi_{f,K}(t) \to \infty \text{ as } t \to \infty$$

and

$$f(t,x,u) \geq \psi_{f,K}(|u|)|u| - a_{f,K}$$

A.J. Zaslavski, *Structure of Approximate Solutions of Optimal Control Problems*, SpringerBriefs in Optimization, DOI 10.1007/978-3-319-01240-7_4, © Alexander J. Zaslavski 2013

for each $t \in R^1$, each $u \in R^m$, and each $x \in R^n$ satisfying $|x| \leq K$;

D(iii) for each $M, \epsilon > 0$ there exist $\Gamma, \delta > 0$ such that

$$|f(t, x_1, u_1) - f(t, x_2, u_2)| \leq \epsilon \max\{f(t, x_1, u_1), f(t, x_2, u_2)\}$$

for each $t \in R^1$, each $u_1, u_2 \in R^m$ and each $x_1, x_2 \in R^n$ which satisfy

$$|x_i| \leq M, \ |u_i| \geq \Gamma, \ i = 1, 2, \quad \max\{|x_1 - x_2|, |u_1 - u_2|\} \leq \delta;$$

D(iv) for each $M, \epsilon > 0$ there exists $\delta > 0$ such that

$$|f(t, x_1, u_1) - f(t, x_2, u_2)| \leq \epsilon$$

for each $t \in R^1$, each $u_1, u_2 \in R^m$, and each $x_1, x_2 \in R^n$ which satisfy

$$|x_i|, |u_i| \leq M, \ i = 1, 2, \quad \max\{|x_1 - x_2|, |u_1 - u_2|\} \leq \delta.$$

D(v) the function f is bounded on $R^1 \times E$ for any bounded set $E \subset R^n \times R^m$.

Note that assumption D(ii) implies that the function f grows superlinearly with respect to u while assumption D(iv) means the uniform continuity of the function with respect to x and u on bounded sets.

It is an elementary exercise to show that an integrand $f = f(t, x, u) \in C^1(R^{n+m+1})$ belongs to $\mathfrak{M}_c$ if f satisfies Assumptions A and D(i), there exists a constant $a_f > 0$ and an increasing function

$$\psi_f : [0, \infty) \to [0, \infty)$$

such that

$$\psi_f(t) \to \infty \text{ as } t \to \infty$$

and

$$f(t, x, u) \geq \psi_f(|u|)|u| - a_f$$

for each $t \in R^1$, each $u \in R^m$ and each $x \in R^n$,

$$\sup\{|f(t, 0, 0)| : t \in R^1\} < \infty$$

and there exists an increasing function $\psi_2 : [0, \infty) \to [0, \infty)$ such that

$$\max\{|\partial f/\partial x(t, x, u)|, \ |\partial f/\partial u(t, x, u)|\} \leq \psi_2(|x|)(1 + \psi(|u|)|u|)$$

for each $t \in R^1$, each $x \in R^n$ and each $u \in R^m$.

We can easily deduce the following result (for the proof see Proposition 2.1 of [48]).

Proposition 4.1. *Let $f \in \mathfrak{M}_c$ and M, ϵ be positive numbers. Then there exist $\Gamma > 0$ and $\delta > 0$ such that*

$$|f(t, x_1, u_1) - f(t, x_2, u_2)| \leq \epsilon \min\{f(t, x_1, u_1), f(t, x_2, u_2)\} \quad (4.4)$$

for each $t \in R^1$, each $u_1, u_2 \in R^m$, and each $x_1, x_2 \in R^n$ which satisfy

$$|x_i| \leq M, \ |u_i| \geq \Gamma, \ i = 1, 2, \quad \max\{|x_1 - x_2|, |u_1 - u_2|\} \leq \delta. \quad (4.5)$$

In the next section we prove the following result.

Proposition 4.2. *Let $f \in \mathfrak{M}_c$. Then f has (LSC) property.*

4.2 Proof of Proposition 4.2

Let $-\infty < T_1 < T_2 < \infty$ and

$$x_j : [T_1, T_2] \to R^n, \ u_j : [T_1, T_2] \to R^m, \ j = 1, 2, \ldots$$

be a sequence of trajectory-control pairs which satisfies

$$\sup\{I^f(T_1, T_2, x_j, u_j) : \ j = 1, 2, \ldots\} < \infty. \quad (4.6)$$

By Proposition 2.7 there exists a number $M_0 > 0$ such that

$$|x_i(t)| \leq M_0, \ t \in [T_1, T_2], \ i = 1, 2, \ldots. \quad (4.7)$$

It follows from (4.6), assumption D(ii) and the properties of the functions ψ_{f, M_0} that the sequence of functions $\{|u_i(\cdot)|\}_{i=1}^{\infty}$ is equiabsolutely integrable on $[T_1, T_2]$. Therefore there exists subsequence $\{(x_{i_k}, u_{i_k})_{k=1}^{\infty}$ and

$$h_1 \in L^1(R^n; (T_1, T_2)), \ h_2 \in L^1(R^m; (T_1, T_2))$$

such that

$$x_{i_k} \to h_1 \text{ as } k \to \infty \text{ weakly in } L^1(R^n; (T_1, T_2)),$$

$$u_{i_k} \to h_2 \text{ as } k \to \infty \text{ weakly in } L^1(R^m; (T_1, T_2)). \quad (4.8)$$

This implies that

$$Ax_{i_k} + Bu_{i_k} \to Ah_1 + Bh_2 \text{ as } k \to \infty \text{ weakly in } L^1(R^n; (T_1, T_2)). \quad (4.9)$$

We may assume without loss of generality that there exists

$$\lim_{k \to \infty} x_{i_k}(T_1).$$

For each $t \in [T_1, T_2]$ set

$$x(t) = \lim_{k \to \infty} x_{i_k}(T_1) + \int_{T_1}^{t} [Ah_1(t) + Bh_2(t)]dt. \tag{4.10}$$

Clearly,

$$x_{i_k}(t) \to x(t) \text{ as } k \to \infty$$

for any $t \in [T_1, T_2]$ and

$$h_1 = x.$$

By Proposition 2.8,

$$x_{i_k}(t) \to x(t) \text{ as } k \to \infty \text{ uniformly in } [T_1, T_2]. \tag{4.11}$$

It remains to show that

$$I^f(T_1, T_2, x, h_2) \leq \liminf_{k \to \infty} I^f(T_1, T_2, x_{i_k}, u_{i_k}).$$

We show that

$$I^f(T_1, T_2, x_{i_k}, u_{i_k}) - \int_{T_1}^{T_2} f(t, x(t), u_{i_k}(t))dt \to 0 \text{ as } k \to \infty. \tag{4.12}$$

Set

$$M_1 = \sup\{I^f(T_1, T_2, x_j, u_j) : \ j = 1, 2, \dots\} < \infty,$$
$$M_2 = \liminf_{k \to \infty}\{I^f(T_1, T_2, x_{i_k}, u_{i_k}). \tag{4.13}$$

Let $\epsilon > 0$. By (4.7) and (4.11),

$$|x(t)| \leq M_0, \ t \in [T_1, T_2]. \tag{4.14}$$

Fix a positive number

$$\gamma < 4^{-1}[1 + M_1 + a_0(T_2 - T_1)]^{-1}\epsilon. \tag{4.15}$$

By Proposition 4.1 there exist $\Gamma > 0$ and $\delta_0 > 0$ such that

$$|f(t, y_1, u_1) - f(t, y_2, u_2)| \leq \gamma \min\{f(t, y_1, u_1), f(t, y_2, u_2)\} \qquad (4.16)$$

for each $t \in R^1$, each $u_1, u_2 \in R^m$, and each $y_1, y_2 \in R^n$ which satisfy

$$|y_i| \leq M_0, \ |u_i| \geq \Gamma, \ i = 1, 2, \quad \max\{|y_1 - y_2|, |u_1 - u_2|\} \leq \delta_0. \qquad (4.17)$$

Since the function f is continuous there exists a number

$$\delta \in (0, \delta_0) \qquad (4.18)$$

such that

$$|f(t, y_1, u_1) - f(t, y_2, u_2)| \leq [8(T_2 - T_1 + 1)]^{-1}\epsilon \qquad (4.19)$$

for each $t \in [T_1, T_2]$, each $u_1, u_2 \in R^m$, and each $y_1, y_2 \in R^n$ which satisfy

$$|y_i| + |u_i| \leq 2M_0 + 2\Gamma + 2, \ i = 1, 2,$$
$$|y_1 - y_2| + |u_1 - u_2| \leq \delta. \qquad (4.20)$$

By (4.11) there exists an integer $k_0 \geq 1$ such that for every integer $k \geq k_0$,

$$|x_{i_k}(t) - x(t)| \leq 2^{-1}\delta, \ t \in [T_1, T_2]. \qquad (4.21)$$

Fix an integer $k \geq k_0$. Set

$$E_1 = [t \in [T_1, T_2] : \ |u_{i_k}(t)| \geq \Gamma\},$$
$$E_2 = [T_1, T_2] \setminus E_1. \qquad (4.22)$$

Clearly,

$$I^f(T_1, T_2, x_{i_k}, u_{i_k}) - \int_{T_1}^{T_2} f(t, x(t), u_{i_k}(t))dt = \sigma_1 + \sigma_2, \qquad (4.23)$$

where

$$\sigma_j = \int_{E_j} [f(t, x_{i_k}(t), u_{i_k}(t)) - f(t, x(t), u_{i_k}(t))]dt, \ j = 1, 2. \qquad (4.24)$$

We estimate σ_1, σ_2 separately.

Let $t \in E_2$. It follows from (4.22), (4.7), (4.14), (4.21), and the choice of δ (see (4.19)) that

$$|f(t, x_{i_k}(t), u_{i_k}(t)) - f(t, x(t), u_{i_k}(t))| \leq [8(T_2 - T_1 + 1)]^{-1}\epsilon.$$

This implies that

$$|\sigma_2| \le 8^{-1}\epsilon. \tag{4.25}$$

Let $t \in E_1$. It follows from (4.22), (4.7), (4.14), (4.18), (4.21), and the choice of Γ, δ_0 (see (4.17)) that

$$|f(t, x_{i_k}(t), u_{i_k}(t)) - f(t, x(t), u_{i_k}(t))|$$
$$\le \gamma f(t, x_{i_k}(t), u_{i_k}(t)).$$

Together with assumption (A), (4.13) and (4.15) this implies that

$$|\sigma_1| \le \gamma \int_{E_1} f(t, x_{i_k}(t), u_{i_k}(t))dt$$
$$\le \gamma(I^f(T_1, T_2, x_{i_k}, u_{i_k}) + a_0(T_2 - T_1))$$
$$\le \gamma(M_1 + a_0(T_2 - T_1)) \le 4^{-1}\epsilon. \tag{4.26}$$

Combining (4.23)–(4.26) we obtain that

$$|I^f(T_1, T_2, x_{i_k}, u_{i_k}) - \int_{T_1}^{T_2} f(t, x(t), u_{i_k}(t))dt| \le \epsilon$$

for each integer $k \ge k_0$. Therefore (4.12) is valid.

We show that

$$I^f(T_1, T_2, x, h_2) \le M_2. \tag{4.27}$$

Let $\delta > 0$. We may assume without loss of generality that

$$I^f(T_1, T_2, x_{i_k}, u_{i_k}) \le M_2 + 2^{-1}\delta \text{ for each integer } k \ge 1. \tag{4.28}$$

Denote by $\mathcal{F}$ the set of all functions $u \in L^1(R^m; (T_1, T_2))$ satisfying

$$\int_{T_1}^{T_2} f(t, x(t), u(t))dt \le M_2 + \delta.$$

It follows from assumption D(i) that the set $\mathcal{F}$ is convex. By (4.12) and (4.28),

$$u_{i_k} \in \mathcal{F} \text{ for all large integers } k. \tag{4.29}$$

Since the set $\mathcal{F}$ is convex it follows (4.29) and (4.8) that there exists a sequence $\{v_i\}_{i=1}^{\infty} \subset \mathcal{F}$ which satisfies

$$\int_{T_1}^{T_2} |v_i(t) - h_2(t)|dt \to 0 \text{ as } i \to \infty.$$

We can assume by extracting a subsequence and re-indexing that

$$v_i(t) \to h_2(t) \text{ as } i \to \infty \text{ a. e. in } [T_1, T_2].$$

It follows from this relation, the continuity of f, assumption (A), the definition of $\mathcal{F}$, and Fatau's lemma that

$$\int_{T_1}^{T_2} f(t, x(t), h_2(t)) dt$$

$$\leq \sup \left\{ \int_{T_1}^{T_2} f(t, x(t), v_i(t)) dt : i = 1, 2, \ldots \right\} \leq M_2 + \delta$$

and

$$\int_{T_1}^{T_2} f(t, x(t), h_2(t)) dt \leq M_2 + \delta.$$

Since this relation holds for any positive δ we conclude that

$$\int_{T_1}^{T_2} f(t, x(t), h_2(t)) dt \leq M_2.$$

This completes the proof of Proposition 4.2.

4.3   A Continuity Property

We use the following result [25].

Proposition 4.3. *For every* $\tilde{y}, \tilde{z} \in R^n$ *and every* $T > 0$ *there exists a unique solution* $x(\cdot)$, $y(\cdot)$ *of the following system*

$$x' = Ax + BB^t y,$$

$$y' = x - A^t y$$

with the boundary constraints

$$x(0) = \tilde{y}, \ x(T) = \tilde{z}$$

(where B^t *is a transpose of* B*).*

Corollary 4.4. *Let* $f \in \mathfrak{M}_c$, $-\infty < T_1 < T_2 < \infty$ *and* $x, y \in R^n$*. Then* $U^f(T_1, T_2, x, y) < \infty$.

Let τ be a positive number. It follows from Proposition 4.3 that for each $\tilde{y}$, $\tilde{z} \in R^n$ there exists a unique solution $x(\cdot)$, $y(\cdot)$ of the following system

$$(x', y')^t = C((x, y)^t) \tag{4.30}$$

with the boundary constraints $x(0) = \tilde{y}$, $x(\tau) = \tilde{z}$ and

$$C((x, y)^t) = (Ax + BB^t y, x - A^t y)^t. \tag{4.31}$$

For any initial value $(x_0, y_0) \in R^n \times R^n$ there exists a unique solution of (4.30)

$$(x(s), y(s))^t = e^{sC}(x_0, y_0)^t, \ s \in R^1.$$

Clearly, for each $\tilde{y}$, $\tilde{z} \in R^n$ there exists a unique pair of vectors

$$D_{\tau,1}(\tilde{y}, \tilde{z}), \ D_{\tau,2}(\tilde{y}, \tilde{z}) \in R^n$$

such that the function

$$(x(s), y(s)) = (e^{sC}((D_{\tau,1}(\tilde{y}, \tilde{z}), D_{\tau,2}(\tilde{y}, \tilde{z}))^t)^t, \ s \in R^1 \tag{4.32}$$

satisfies (4.30) with the boundary constraints

$$x(0) = \tilde{y}, \ x(\tau) = \tilde{z}.$$

It is easy to see that

$$D_{\tau,j} : R^n \times R^n \to R^n, \ j = 1, 2$$

are linear operators.

Using assumption D(v) we can easily deduce the following result.

Proposition 4.5. *Let $f \in \mathfrak{M}_c$, $M, \tau > 0$. Then*

$$\sup\{|U^f(T, T + \tau, y, z)| : \ T \in R^1, \ y, z \in R^n, \ |y|, |z| \le M\} < \infty.$$

Proposition 4.6. *Let $f \in \mathfrak{M}_c$, $M, \tau, \epsilon > 0$. Then there exists a number $\delta > 0$ such that:*

1. for each $T \in R^1$, each $y_1, y_2, z_1, z_2 \in R^n$ satisfying

$$|y_i|, |z_i| \le M, \ i = 1, 2, \ |y_1 - y_2|, \ |z_1 - z_2| \le \delta \tag{4.33}$$

the following relation holds:

$$|U^f(T, T + \tau, y_1, z_1) - U^f(T, T + \tau, y_2, z_2)| \le \epsilon. \tag{4.34}$$

2. *for each $T \in R^1$, each $y_1, y_2, z_1.z_2 \in R^n$ satisfying (4.33) and each trajectory-control pair*

$$x_1 : [T, T + \tau] \to R^n, \ u_1 : [T, T + \tau] \to R^m$$

which satisfies

$$x_1(T_1) = y_1, \ x_1(T_2) = z_1,$$
$$I^f(T_1, T_2, x_1, u_1) = U^f(T_1, T_2, y_1, z_1)$$

there exists a trajectory-control pair

$$x_2 : [T, T + \tau] \to R^n, \ u_2 : [T, T + \tau] \to R^m$$

such that

$$x_2(T_1) = y_2, \ x_2(T_2) = z_2,$$
$$|I^f(T_1, T_2, x_2, u_2) - I^f(T_1, T_2, x_1, u_1)| \le \epsilon,$$
$$|x_1(t) - x_2(t)| \le \epsilon, \ t \in [T_1, T_2].$$

Proof. Set

$$M_0 = \sup\{|U^f(T, T + \tau, y, z)| : \ T \in R^1, \ y, z \in R^n, \ |y|, |z| \le M + 4\} < \infty. \tag{4.35}$$

By Proposition 4.5, $M_0 < \infty$. By Proposition 2.7 there exists a number $M_1 > 0$ such that for each $T \in R^1$ and each trajectory-control pair

$$x : [T, T + \tau] \to R^n, \ u : [T, T + \tau] \to R^m$$

satisfying

$$I^f(T, T + \tau, x, u) \le M_0 + 4$$

the following relation holds:

$$|x(t)| \le M_1, \ t \in [T, T + \tau]. \tag{4.36}$$

Choose a number $\delta_1 > 0$ such that

$$4\delta_1(M_0 + a_0\tau + \tau) < \epsilon. \tag{4.37}$$

By Proposition 4.1 there exist $\Gamma_0 > 8$ and $\delta_2 \in (0, 8^{-1})$ such that

$$|f(t, x_1, u_1) - f(t, x_2, u_2)| \leq \delta_1 \min\{f(t, x_1, u_1),\ f(t, x_2, u_2)\} \qquad (4.38)$$

for each $t \in R^1$, each $x_1, x_2 \in R^n$, and each $u_1, u_2 \in R^m$ which satisfy

$$|x_i| \leq M_1 + 8,\ i = 1, 2,\ |u_i| \geq \Gamma_0 - 4,\ i = 1, 2,$$
$$|x_1 - x_2|,\ |u_1 - u_2| \leq \delta_2. \qquad (4.39)$$

By assumption D(iv) there exists a number

$$\delta_3 \in (0, 4^{-1} \min\{\delta_1, \delta_2, \epsilon\}) \qquad (4.40)$$

such that

$$|f(t, x_1, u_1) - f(t, x_2, u_2)| \leq \delta_1 \qquad (4.41)$$

for each $t \in R^1$, each $x_1, x_2 \in R^n$, and each $u_1, u_2 \in R^m$ which satisfy

$$|x_i|,\ |u_i| \leq \Gamma_0 + M_1 + 4,\ i = 1, 2,$$
$$|x_1 - x_2|,\ |u_1 - u_2| \leq \delta_3. \qquad (4.42)$$

There exists a number

$$\delta \in (0, 8^{-1}\delta_3) \qquad (4.43)$$

such that

$$(1 + ||B||)|e^{tC}((D_{\tau,1}(y, z), D_{\tau,2}(y, z))^t)| \leq 2^{-1}\delta_3,\ t \in [0, \tau] \qquad (4.44)$$

for each $y, z \in R^n$ satisfying $|y|, |z| \leq \delta$.

Assume that $T \in R^1$ and $y_1, y_2, z_1.z_2 \in R^n$ satisfy (4.33). By Proposition 4.2 and Corollary 4.4, there exists a trajectory-control pair

$$x_1 : [T, T + \tau] \to R^n,\ u_1 : [T, T + \tau] \to R^m$$

such that

$$x_1(T) = y_1,\ x_1(T + \tau) = z_1,$$
$$I^f(T, T + \tau, x_1, u_1) = U^f(T, T + \tau, y_1, z_1). \qquad (4.45)$$

It follows from (4.45), (4.35), (4.33), and the definition of M_1 that

$$|x_1(t)| \leq M_1, \ t \in [T, T + \tau]. \tag{4.46}$$

Define functions $h_j : R^1 \to R^n, j = 1, 2$ as follows:

$$(h_1(s), h_2(s))^t = e^{sC}((D_{\tau,1}(y_2 - y_1, z_2 - z_1), D_{\tau,2}(y_2 - y_1, z_2 - z_1))^t), \ s \in R^1. \tag{4.47}$$

Set

$$x_2(t) = x_1(t) + h_1(t - T), \ u_2(t) = u_1(t) + B^t h_2(t - T), \ t \in [T, T + \tau]. \tag{4.48}$$

It follows from (4.47), (4.48), (4.45), (4.30)–(4.32), and the definition of $D_{\tau,1}, D_{\tau,2}$ that

$$(x_2, u_2) \text{ is a trajectory-control pair and}$$
$$x_2(T) = y_2, \ x_2(T + \tau) = z_2. \tag{4.49}$$

By (4.48), (4.47), (4.44), (4.33), and the choice of δ,

$$|x_2(t) - x_1(t)|, \ |u_2(t) - u_1(t)| \leq 2^{-1}\delta_3, \ t \in [T, T + \tau]. \tag{4.50}$$

Assumption (A), (4.45), (4.35), and (4.33) imply that for any measurable set $E \subset [T, T + \tau]$,

$$\int_E f(t, x_1(t), u_1(t))dt \leq M_0 + a_0\tau. \tag{4.51}$$

Set

$$E_1 = \{t \in [T, T + \tau] : |u_1(t)| \geq \Gamma_0\},$$
$$E_2 = [T, T + \tau] \setminus E_1.$$

It follows from (4.46), (4.50), (4.40), and the choice of Γ_0, δ_2 and (4.51) that

$$\int_{E_1} |f(t, x_1(t), u_1(t)) - f(t, x_2(t), u_2(t))|dt$$
$$\leq \delta_1 \int_{E_1} f(t, x_1(t), u_1(t))dt \leq \delta_1(M_0 + a_0\tau). \tag{4.52}$$

By (4.46), (4.50) and the choice of δ_3,

$$\int_{E_2} |f(t, x_1(t), u_1(t)) - f(t, x_2(t), u_2(t))|dt \leq \delta_1\tau.$$

This relation, (4.52), (4.49), (4.45), and (4.37) imply that

$$|I^f(T_1, T_2, x_2, u_2) - I^f(T_1, T_2, x_1, u_1)| \leq \epsilon,$$

$$U^f(T, T + \tau, y_2, z_2) \leq U^f(T, T + \tau, y_1, z_1) + \epsilon.$$

This completes the proof of Proposition 4.6.

4.4 A Boundedness Property

We can easily deduce the following result.

Proposition 4.7. *Assume that* $f \in \mathfrak{M}_c$, $\tau, M_1, M_2 > 0$ *and that*

$$\inf\{U^f(T, T + \tau, x, y) : T \in R^1, \ x, y \in R^n, \ |x| + |y| \geq M_1\}$$

$$> \sup\{|U^f(T, T + \tau, 0, 0)| : T \in R^1\} + 1. \tag{4.53}$$

Then there exists an integer $N > 2$ *such that:*

1. *For each* $T \in R^1$, *each integer* $q \geq N$ *and each sequence* $\{x_k\}_{k=0}^q \subset R^n$
   satisfying

$$\{k \in \{0, \ldots, q\} : |x_k| \leq M_1\} = \{0, q\}$$

   the following relation holds:

$$\sum_{k=0}^{q-1} \left[U^f(T+k\tau, T+(k+1)\tau, x_k, x_{k+1}) - U^f(T+k\tau, T+(k+1)\tau, y_k, y_{k+1}) \right] \geq M_2, \tag{4.54}$$

   where

$$y_k = x_k, \ k = 0, q, \ y_k = 0, \ k = 1, \ldots, q - 1.$$

2. *For each* $T \in R^1$, *each integer* $q \geq N$, *and each sequence* $\{x_k\}_{k=0}^q \subset R^n$
   satisfying

$$\{k \in \{0, \ldots, q\} : |x_k| \leq M_1\} = \{0\}$$

   relation (4.54) holds with $y_0 = x_0$, $y_k = 0$, $k = 1, \ldots, q$.
3. *For each* $T \in R^1$, *each integer* $q \geq N$, *and each sequence* $\{x_k\}_{k=0}^q \subset R^n$
   satisfying

$$\{k \in \{0, \ldots, q\} : |x_k| \leq M_1\} = \{q\}$$

   relation (4.54) holds with $y_q = x_q$, $y_k = 0$, $k = 0, \ldots, q - 1$.

Proposition 4.8. *Assume that $f \in \mathfrak{M}_c$, $\tau, M_1, M_2 > 0$ and (4.53) holds. Then there exists a number $M_3 > M_1 + M_2$ such that:*

1. *For each $T \in R^1$, each integer $q \geq 1$, and each sequence $\{x_k\}_{k=0}^q \subset R^n$ satisfying*

$$\max\{|x_0|, |x_q|\} \leq M_1, \ \max\{|x_k| : \ k = 0, \dots, q\} > M_3 \qquad (4.55)$$

there is a sequence $\{y_k\}_{k=0}^q \subset R^n$ such that $y_0 = x_0$, $y_q = x_q$ and (4.54) holds.

2. *For each $T \in R^1$, each integer $q \geq 1$, and each sequence $\{x_k\}_{k=0}^q \subset R^n$ satisfying*

$$|x_0| \leq M_1, \ \max\{|x_k| : \ k = 0, \dots, q\} > M_3 \qquad (4.56)$$

there is a sequence $\{y_k\}_{k=0}^q \subset R^n$ such that $y_0 = x_0$ and (4.54) is valid.

3. *For each $T \in R^1$, each integer $q \geq 1$, and each sequence $\{x_k\}_{k=0}^q \subset R^n$ satisfying*

$$|x_q| \leq M_1, \ \max\{|x_k| : \ k = 0, \dots, q\} > M_3$$

there is a sequence $\{y_k\}_{k=0}^q \subset R^n$ such that $y_q = x_q$ and (4.54) is valid.

Proof. Let an integer $N > 6$ be as guaranteed in Proposition 4.7. Fix a large number $M_3 > M_1 + M_2$.

We will prove assertion 1. Assume that $T \in R^1$, an integer $q \geq 1$, and a sequence $\{x_k\}_{k=0}^q \subset R^n$ satisfies (4.55). Then there is $j \in \{0, \dots, q\}$ such that $|x_j| > M_3$. Set

$$i_1 = \max\{k \in \{0, \dots, j\} : \ |x_k| \leq M_1\}, \ i_2 = \min\{k \in \{j, \dots, q\} : \ |x_k| \leq M_1\}.$$

If $i_2 - i_1 \geq N$, then by the validity of assertion 1 follows from the definition of N and Proposition 4.7.

If $i_2 - i_1 < N$, then we set

$$y_k = x_k, \ k \in \{0, \dots, i_1\} \cup \{i_2, \dots, q\}, \ y_k = 0, \ k = i_1 + 1, \dots, i_2 - 1 \quad (4.57)$$

and it is easy to see that (4.54) holds when the constant M_3 is large enough.

We will prove assertion 2. Assume that $T \in R^1$, and integer $q \geq 1$ and a sequence $\{x_k\}_{k=0}^q \subset R^n$ satisfies (4.56). Then there is $j \in \{1, \dots, q\}$ such that $|x_j| > M_3$. Set

$$i_1 = \max\{k \in \{0, \dots, j\} : \ |x_k| \leq M_1\}.$$

If $|x_k| > M_1$ for $k = j, \ldots, q$, then we set

$$y_i = x_i, \ i = 0, \ldots, i_1, \ y_i = 0, \ i = i_1 + 1, \ldots, q.$$

Otherwise we set

$$i_2 = \min\{k \in \{j, \ldots, q\} : \ |x_k| \le M_1\}$$

and define $\{y_k\}_{k=0}^{q}$ by (4.57). It is easy to verify that in both cases (4.54) holds.

We will prove assertion 3. Assume that $T \in R^1$, an integer $q \ge 1$, and a sequence $\{x_k\}_{k=0}^{q} \subset R^n$ satisfies

$$|x_q| \le M_1, \ \max\{|x_k| : \ k = 0, \ldots, q\} > M_3.$$

Then there is $j \in \{0, \ldots, q-1\}$ such that $|x_j| > M_3$. Set

$$i_2 = \min\{k \in \{j, \ldots, q\} : \ |x_k| \le M_1\}.$$

If $|x_k| > M_1$ for $k = 0, \ldots, j$, then we set

$$y_i = x_i, \ i = i_2, \ldots, q, \ y_i = 0, \ i = 0, \ldots, i_2 - 1.$$

Otherwise we set

$$i_1 = \max\{k \in \{0, \ldots, j\} : \ |x_k| \le M_1\}$$

and define $\{y_k\}_{k=0}^{q}$ by (4.57). It is easy to verify that in both cases (4.54) holds. This completes the proof of Proposition 4.8.

4.5   The Existence and Structure of Solutions

Proposition 4.9. *Let $f \in \mathfrak{M}_c$. Then f satisfies assumption (B).*

Proof. By Propositions 4.2 and 4.5 and (LSC) property for each integer $N \ge 1$ there exists a trajectory-control pair

$$x_N : [-N, N] \to R^n, \ u_N : [-N, N] \to R^m$$

such that

$$I^f(-N, N, x_N, u_N) \le U^f(-N, N, y, z) \text{ for each } y, z \in R^n. \tag{4.58}$$

Then there exists a number $M_1 > 0$ which satisfies (4.53) with $\tau = 1$. Let a number $M_3 > M_1 + 1$ be as guaranteed in Proposition 4.8 with $\tau = 1$, $M_2 = 1$. We will show that

$$\sup\{|x_N(i)| : \text{ an integer } i \in [-N, N], \ N = 1, 2, \ldots\} \leq M_3. \tag{4.59}$$

Let $N \geq 1$ be an integer. By (4.58) and (4.53) which holds with $\tau = 1$ there exists an integer $q \in [-N, N]$ such that

$$|x_N(q)| \leq M_1.$$

Relation (4.59) follows from this relation, (4.58), Proposition 4.8 which holds with $\tau = 1$, $M_2 = 1$ and the choice of M_3.

By (4.58), (4.59) and Propositions 2.7 and 4.5,

$$\sup\{|x_N(t)| : t \in [-N, N], \ N = 1, 2, \ldots\} < \infty. \tag{4.60}$$

By (4.58), (4.60), and Proposition 4.5 there exists a number $M_4 > 0$ such that

$$I^f(-i, i, x_p, u_p) \leq I^f(-i, i, x_i, u_i) + M_4 \tag{4.61}$$

for each pair of integers $i \geq 1$ and $p \geq i$. Together with assumption (A) this implies that for each integer j the sequence

$$\{I^f(j, j+1, x_p, u_p) : p \geq |j| + 1\}$$

is bounded. Therefore by Proposition 4.2 and (LSC) property there exists a subsequence $\{(x_{N_k}, u_{N_k})\}_{k=1}^{\infty}$ and a trajectory-control pair

$$x_f : R^1 \to R^n, \ u_f : R^1 \to R^m$$

such that

$$x_{N_k}(t) \to x_f(t) \text{ as } k \to \infty \text{ for any } t \in R^1 \tag{4.62}$$

and

$$I^f(j, j+1, x_f, u_f) \leq \liminf_{k \to \infty} I^f(j, j+1, x_{N_k}, u_{N_k}). \tag{4.63}$$

It follows from (4.60)–(4.63) that

$$\sup\{|x_f(t)| : t \in R^1\} < \infty$$

and

$$I^f(-i, i, x_f, u_f) \leq I^f(-i, i, x_i, u_i) + M_4 \tag{4.64}$$

for each integer $i \geq 1$.

By (4.63), (4.58), (4.62), and Proposition 4.6,

$$I^f(i, j, x_f, u_f) \leq \liminf_{k \to \infty} I^f(i, j, x_{N_k}, u_{N_k})$$

$$\leq \liminf_{k \to \infty} U^f(i, j, x_{N_k}(i), x_{N_k}(j)) = U^f(i, j, x_f(i), x_f(j)). \tag{4.65}$$

Therefore f satisfies assumption B(i). It follows from (4.65), (4.64), and Proposition 4.5 that f satisfies assumption B(ii).

We show that f satisfies assumption B(iii). Let $S_1 > 0$. Set $c = 1$ and

$$c_0 = \sup\{|x_f(t)| : t \in R^1\} + S_1 + 2. \tag{4.66}$$

By Proposition 4.5 there exists a number

$$c_1 > \sup\{U^f(T, T + 1, y, z) :$$

$$T \in R^1, \; y, z \in R^n, \; |y|, |z| \leq c_0 + 1\}. \tag{4.67}$$

Choose a number

$$S_2 > M_4 + 2a_0 + 2c_1 + 4 \tag{4.68}$$

(recall a_0 in assumption (A)).

Suppose that $T_1 \in R^1$, $T_2 \geq T_1 + 1$ and a trajectory-control pair

$$x : [T_1, T_2] \to R^n, \; u : [T_1, T_2] \to R^m$$

satisfies

$$|x(T_i)| \leq S_1, \; i = 1, 2. \tag{4.69}$$

Choose an integer

$$N > |T_1| + |T_2| + 4. \tag{4.70}$$

By Proposition 4.5 there exists trajectory-control pair

$$y : [-N, N] \to R^n, \; u : [-N, N] \to R^m$$

such that

$$y(t) = x_f(t), \; v(t) = u_f(t), \; t \in [-N, T_1 - 1] \cup [T_2 + 1, N],$$
$$y(t) = x(t), \; v(t) = u(t), \; t \in [T_1, T_2],$$
$$I^f(\tau, \tau + 1, y, v) = U^f(\tau, \tau + 1, y(\tau), y(\tau + 1)) + 1, \; \tau = T_1 - 1, \, T_2. \quad (4.71)$$

It follows from (4.71), (4.64), and (4.58) that

$$I^f(-N, N, x_f, u_f) - I^f(-N, N, y, v)$$
$$\leq M_4 + I^f(-N, N, x_N, u_N) - I^f(-N, N, y, v) \leq M_4. \quad (4.72)$$

On the other hand by (4.71), assumption (A), (4.69), (4.66), (4.67),

$$I^f(-N, N, x_f, u_f) - I^f(-N, N, y, v)$$
$$\geq I^f(T_1, T_2, x_f, u_f) - I^f(T_1, T_2, x, u) - 2a_0 - 2c_1 - 2.$$

Together with (4.72) and (4.68) this implies that

$$I^f(T_1, T_2, x_f, u_f) \leq I^f(T_1, T_2, x, u) + S_2.$$

Therefore f satisfies assumption B(iii). Set $b_f = 1$. It follows from assumption B(i), Proposition 4.6, and (4.64) that f satisfies assumption B(iv). Proposition 4.9 is proved.

Proposition 4.10. $\mathfrak{M}_c$ *is a closed subset of* $\mathfrak{M}$.

Proof. Assume that $f_i \in \mathfrak{M}_c$, $i = 1, 2, \ldots$ and $f_i \to f \in \mathfrak{M}$ as $i \to \infty$. It is sufficient to show that $f \in \mathfrak{M}_c$. Clearly, f is a continuous function. We show that f satisfies assumption (D). It is easy to see that f satisfies assumptions D(i), D(iv), and D(v). We show that f satisfies assumption D(ii).

Let $K > 0$. There exists an integer $j \geq 1$ such that

$$(f, f_j) \in E(K, 2^{-1}, 2)$$

(see (2.7)). Therefore

$$|f(t, x, u)| + 1 \geq 2^{-1}|f_j(t, x, u)| + 2^{-1}, \; (t, x, u) \in R^{n+m+1}, \; |x| \leq K. \quad (4.73)$$

Since $f_j \in \mathfrak{M}_c$ satisfies assumption D(ii) there exists a constant $a_{k,j} > 0$ and an increasing function

$$\psi_{k,j} : [0, \infty) \to [0, \infty)$$

such that

$$\psi_{k,j}(t) \to \infty \text{ as } t \to \infty,$$

$$f_j(t, x, u) \geq \psi_{k,j}(|u|)|u| - a_{k,j}, \ (t, x, u) \in R^{n+m+1}, \ |x| \leq K. \tag{4.74}$$

It follows from (4.73) and (4.74) that

$$|f(t, x, u)| \geq 2^{-1}\psi_{k,j}(|u|)|u| - 2^{-1}a_{k,j} - 2^{-1}, \ (t, x, u) \in R^{n+m+1}, \ |x| \leq K. \tag{4.75}$$

There exists $K_0 \geq 1$ such that

$$\psi_{k,j}(K_0) \geq 8a_0 + 8a_{k,j} + 8. \tag{4.76}$$

By (4.75), (4.76), and assumption (A) for each $(t, x, u) \in R^{n+m+1}$ satisfying $|x| \leq K$, $|u| \geq K_0$,

$$f(t, x, u) \geq 0 \text{ and } f(t, x, u) \geq 2^{-1}\psi_{k,j}(|u|)|u| - a_{k,j} - 1.$$

It is easy to see that for each $(t, x, u) \in R^{n+m+1}$ satisfying $|x| \leq K$,

$$f(t, x, u) \geq 2^{-1}\psi_{k,j}(|u|)|u| - a_{k,j} - 1 - \psi_{k,j}(|K_0|)|K_0| - a_0.$$

Therefore f satisfies assumption D(ii).

We show that f satisfies assumption D(iii).

Let $M, \epsilon > 0$. Fix a number $\lambda > 1$ such that

$$\lambda^2 - 1 < 8^{-1}\epsilon. \tag{4.77}$$

Clearly, there exists an integer $j \geq 1$ such that

$$(f, f_i) \in E(M, \epsilon, \lambda) \text{ for each integer } i \geq j. \tag{4.78}$$

Since f, f_j satisfy assumption D(ii) there exist a number $a_M > 0$ and an increasing function

$$\psi_M : [0, \infty) \to [0, \infty)$$

such that

$$\psi_M(t) \to \infty \text{ as } t \to \infty,$$

$$f_j(t, x, u), \ f(t, x, u) \geq \psi_M(|u|)|u| - a_M, \ (t, x, u) \in R^{n+m+1}, \ |x| \leq M. \tag{4.79}$$

By (4.77) and the properties of ψ_M there exists a number Γ_0 such that

$$\Gamma_0 > 1,$$
$$\psi_M(\Gamma_0) \geq 2a_0 + 2a_M,$$
$$\lambda^2(1 + 2\psi_M(\Gamma_0)^{-1})^2 - 1 < 8^{-1}\epsilon. \tag{4.80}$$

Fix a positive number ϵ_1 which satisfies

$$8\epsilon_1[\lambda(1 + 2\psi_M(\Gamma_0)^{-1})]^2 < \epsilon. \tag{4.81}$$

By Proposition 4.1 there exist numbers $\Gamma, \delta > 0$ such that

$$\Gamma > \Gamma_0,$$
$$|f_j(t, x_1, u_1) - f_j(t, x_2, u_2)| \leq \epsilon_1 \min\{f_j(t, x_1, u_1), f_j(t, x_2, u_2)\} \tag{4.82}$$

for each $t \in R^1$, each $u_1, u_2 \in R^m$, and each $x_1, x_2 \in R^n$ which satisfy

$$|x_i| \leq M, \ |u_i| \geq \Gamma, \ i = 1, 2, \quad \max\{|x_1 - x_2|, |u_1 - u_2|\} \leq \delta. \tag{4.83}$$

Assume that $t \in R^1$ and $u_1, u_2 \in R^m$, $x_1, x_2 \in R^n$ satisfy (4.83). It follows from the choice of Γ, δ that (4.82) holds. By (4.78) and (4.83),

$$(|f(t, x_i, u_i)| + 1)(|f_j(t, x_i, u_i)| + 1)^{-1} \in [\lambda^{-1}, \lambda], \ i = 1, 2. \tag{4.84}$$

Relations (4.83), (4.79), and (4.80) imply that

$$\min\{f_j(t, x_i, u_i), \ f(t, x_i, u_i)\} \geq 2^{-1}\psi_M(\gamma_0), \ i = 1, 2.$$

Together with (4.84) this implies that

$$f(t, x_i, u_i)f_j(t, x_i, u_i)^{-1}$$
$$\in [\lambda(1 + 2\psi_M(\Gamma_0)^{-1})^{-1}, \ \lambda(1 + 2\psi_M(\Gamma_0)^{-1})], \ i = 1, 2. \tag{4.85}$$

We may assume without loss of generality that

$$f(t, x_1, u_1) \geq f(t, x_2, u_2). \tag{4.86}$$

It follows from (4.85), (4.82), (4.81) and (4.80) that

$$f(t, x_1, u_1) - f(t, x_2, u_2)$$
$$\leq \lambda(1 + 2\psi_M(\Gamma_0)^{-1})f_j(t, x_1, u_1) - (\lambda(1 + 2\psi_M(\Gamma_0)^{-1}))^{-1}f_j(t, x_2, u_2)$$
$$= \lambda(1 + 2\psi_M(\Gamma_0)^{-1})[f_j(t, x_1, u_1) - f_j(t, x_2, u_2)]$$

$$+f_j(t, x_2, u_2)[\lambda(1 + 2\psi_M(\Gamma_0)^{-1}) - (\lambda(1 + 2\psi_M(\Gamma_0)^{-1}))^{-1}]$$

$$\leq \epsilon_1\lambda(1 + 2\psi_M(\Gamma_0)^{-1})f_j(t, x_2, u_2)$$

$$+f_j(t, x_2, u_2)[\lambda(1 + 2\psi_M(\Gamma_0)^{-1}) - (\lambda(1 + 2\psi_M(\Gamma_0)^{-1}))^{-1}]$$

$$\leq \epsilon_1[\lambda(1 + 2\psi_M(\Gamma_0)^{-1})]^2 f_j(t, x_2, u_2)$$

$$+f(t, x_2, u_2)[\lambda^2(1 + 2\psi_M(\Gamma_0)^{-1})^2 - 1] \leq \epsilon f(t, x_2, u_2).$$

Therefore the function f satisfies assumption D(iii). This completes the proof of Proposition 4.10.

Assume that $f \in \mathfrak{M}_c$. By Proposition 4.9 f satisfies assumption (B). Let a trajectory-control pair

$$x_f : R^1 \to R^n, \ u_f : R^1 \to R^m$$

be as guaranteed by assumption (B).

For each $r > 0$ we define $f_r : R^{n+m+1} \to R^1$ as follows:

$$f_r(t, x, u) = f(t, x, u) + r\min\{|x - x_f(t)|, 1\}, \ (t, x, u) \in R^{n+m+1}.$$

It is easy to see that $f_r \in \mathfrak{M}_c$ for each $r > 0$.

By Proposition 4.10 $\mathfrak{M}_c$ is a closed subset of $\mathfrak{M}$. It follows from Proposition 4.9 that $\mathfrak{M}_c \subset \mathfrak{M}_{reg}$. Moreover $f_r \in \mathfrak{M}_c$ for each $f \in \mathfrak{M}_c$ and each $r > 0$.

It follows from the results of Chap. 2 that there exists a set $\mathcal{F} \subset \mathfrak{M}_c$ which is a countable intersection of open everywhere dense sets in $\mathfrak{M}_c$ and for which Theorems 2.2, 2.3 and 2.5 hold.

Theorem 3.5 and Propositions 4.2 and 4.9 imply the following result.

Theorem 4.11. *Assume that $f \in \mathfrak{M}_c$ has the turnpike property, $\tau \in R^1$, and*

$$x : [\tau, \infty) \to R^n, \ u : [\tau, \infty) \to R^m$$

is an (f)-good trajectory-control pair. Then there exists an (f)-overtaking optimal-trajectory pair

$$x_* : [\tau, \infty) \to R^n, \ u_* : [\tau, \infty) \to R^m$$

such that $x_*(\tau) = x(\tau)$.

References

1. Anderson, B.D.O., Moore, J.B.: Linear Optimal Control. Prentice-Hall, Englewood Cliffs (1971)
2. Arkin, V.I., Evstigneev, I.V.: Stochastic Models of Control and Economic Dynamics. Academic, London (1987)
3. Aseev, S.M., Besov, K.O., Kryazhimskii, A.V.: Russ. Math. Surv. **67**, 195–253 (2012)
4. Aseev, S.M., Kryazhimskiy, A.V.: SIAM J. Control Optim. **43**, 1094–1119 (2004)
5. Aseev, S.M., Veliov, V.M.: Dyn. Contin., Discret. Impulsive Syst. Ser. B: Appl. Algorithms **19**, 43–63 (2012)
6. Aubin, J.P., Ekeland, I.: Applied Nonlinear Analysis. Wiley Interscience, New York (1984)
7. Aubry, S., Le Daeron, P.Y.: Physica D **8**, 381–422 (1983)
8. Baumeister, J., Leitao, A., Silva, G.N.: Syst. Control Lett. **56**, 188–196 (2007)
9. Blot, J., Cartigny, P.: J. Optim. Theory Appl. **106**, 411–419 (2000)
10. Blot, J., Hayek, N.: ESAIM Control Optim. Calc. Var. **5**, 279–292 (2000)
11. Blot, J., Michel, P.: Appl. Math. Lett. **16**, 71–78 (2003)
12. Carlson, D.A., Haurie, A., Leizarowitz, A.: Infinite Horizon Optimal Control. Springer, Berlin (1991)
13. Cartigny, P., Michel, P.: Automatica J. IFAC **39**, 1007–1010 (2003)
14. Coleman, B.D., Marcus, M., Mizel, V.J.: Arch. Rational Mech. Anal. **117**, 321–347 (1992)
15. Evstigneev, I.V., Flam, S.D.: Set-Valued Anal. **6**, 61–81 (1998)
16. Gaitsgory, V., Rossomakhine, S., Thatcher, N.: Dyn. Contin., Discret. Impulsive Syst. Ser. B: Appl. Algorithms **19**, 43–63 (2012)
17. Gale, D.: Rev. Econ. Stud. **34**, 1–18 (1967)
18. Guo, X., Hernandez-Lerma, O.: Bernoulli **11**, 1009–1029 (2005)
19. Hammond, P.J.: Consistent Planning and Intertemporal Welfare Economics. University of Cambridge, Cambridge (1974)
20. Hammond, P.J.: Rev. Econ. Stud. **42**, 1–14 (1975)
21. Hammond, P.J., Mirrlees, J.A.: Models of Economic Growth, pp. 283–299. Wiley, New York (1973)
22. Hayek, N.: Optimization **60**, 509–529 (2011)
23. Jasso-Fuentes, H., Hernandez-Lerma, O.: Appl. Math. Optim. **57**, 349–369 (2008)
24. Leizarowitz, A.: Appl. Math. Optim. **13**, 19–43 (1985)
25. Leizarowitz, A.: Appl. Math. Optim. **14**, 155–171 (1986)
26. Leizarowitz, A., Mizel, V.J.: Arch. Rational Mech. Anal. **106**, 161–194 (1989)
27. Lykina, V., Pickenhain, S., Wagner, M.: J. Math. Anal. Appl. **340**, 498–510 (2008)
28. Makarov, V.L., Rubinov, A.M.: Mathematical Theory of Economic Dynamics and Equilibria. Springer, New York (1977)

A.J. Zaslavski, *Structure of Approximate Solutions of Optimal Control Problems*,
SpringerBriefs in Optimization, DOI 10.1007/978-3-319-01240-7,
© Alexander J. Zaslavski 2013

29. Malinowska, A.B., Martins, N., Torres, D.F.M.: Optim. Lett. **5**, 41–53 (2011)
30. Marcus, M., Zaslavski, A.J.: Ann. Inst. H. Poincare, Anal. Non Lineare **16**, 593–629 (1999)
31. Marcus, M., Zaslavski, A.J.: Ann. Inst. H. Poincare, Anal. Non Lineare **19**, 343–370 (2002)
32. Martins, N., Torres, D.F.M.: J. Optim. Theory Appl. **155**(2), 453–476 (2012)
33. McKenzie, L.W.: Econometrica **44**, 841–866 (1976)
34. Mordukhovich, B.S.: Automat. Remote Control **50**, 1333–1340 (1990)
35. Mordukhovich, B.S.: Appl. Anal. **90**, 1075–1109 (2011)
36. Mordukhovich, B.S., Shvartsman, I.: Optimal Control, Stabilization and Nonsmooth Analysis. Lecture Notes in Control and Information Sciences, pp. 121–132. Springer, New York (2004)
37. Moser, J.: Ann. Inst. H. Poincare, Anal. Nonlineare **3**, 229–272 (1986)
38. Ocana, E., Cartigny, P.: SIAM J. Control Optim. **50**, 2573–2587 (2012)
39. Ocana, E., Cartigny, P., Loisel, P.: J. Nonlinear Convex Anal. **10**, 157–176 (2009)
40. Pickenhain, S., Lykina, V., Wagner, M.: Control Cybernet **37**, 451–468 (2008)
41. Rubinov, A.M.: J. Soviet Math. **26**, 1975–2012 (1984)
42. Samuelson, P.A.: Am. Econ. Rev. **55**, 486–496 (1965)
43. von Weizsacker, C.C.: Rev. Econ. Stud. **32**, 85–104 (1965)
44. Zaslavski, A.J.: Math. USSR Izvestiya **29**, 323–354 (1987)
45. Zaslavski, A.J.: SIAM J. Control Optim. **33**, 1643–1660 (1995)
46. Zaslavski, A.J.: SIAM J. Control Optim. **33**, 1661–1686 (1995)
47. Zaslavski, A.J.: Nonlinear Anal. **27**, 895–931 (1996)
48. Zaslavski, A.J.: Abstr. Appl. Anal. **3**, 265–292 (1998)
49. Zaslavski, A.J.: Nonlinear Anal. **42**, 1465–1498 (2000)
50. Zaslavski, A.J.: Turnpike Properties in the Calculus of Variations and Optimal Control. Springer, New York (2006)
51. Zaslavski, A.J.: Optimization on Metric and Normed Spaces. Springer, New York (2010)
52. Zaslavski, A.J.: Portugal. Math. **68**, 239–257 (2011)
53. Zaslavski, A.J., Leizarowitz, A.: Math. Oper. Res. **22**, 726–746 (1997)

Index

A
Absolutely continuous function, 5
Agreeable trajectory-control pair, 91
Approximate solution, 4

B
Borelian function, 12
Borel measurable set, 11

C
Complete metric space, 1
Complete uniform space, 3
Control constraint, 11
Control function, 11
Control system, 9

D
Differential equation, 11

E
Euclidean norm, 2
Euclidean space, 2

G
Good function, 4
Good point, 81

I
Increasing function, 2

Infinite horizon problem, 1
Integral functional, 12
Integrand, 2

L
Lebesgue integrable function, 12
Lebesgue measurable function, 12
(LSC) property, 79

M
Minimal solution, 2

N
Neumann path, 6

O
Overtaking optimality criterion, 1

T
Topological subspace, 15
Trajectory-control pair, 12
Turnpike property, 6

U
Uniformity, 3
Uniform space, 3